Advancements in Hydropower Design and Operation for Present and Future Electrical Demand

Advancements in Hydropower Design and Operation for Present and Future Electrical Demand

Editors

John M. Cimbala
Bryan J. Lewis

MDPI • Basel • Beijing • Wuhan • Barcelona • Belgrade • Manchester • Tokyo • Cluj • Tianjin

Editors

John M. Cimbala
The Pennsylvania State University
USA

Bryan J. Lewis
Brigham Young University-Idaho
USA

Editorial Office
MDPI
St. Alban-Anlage 66
4052 Basel, Switzerland

This is a reprint of articles from the Special Issue published online in the open access journal *Energies* (ISSN 1996-1073) (available at: https://www.mdpi.com/journal/energies/special_issues/Advancements_Hydropower_Design_Operation_Present_Future_Electrical_Demand).

For citation purposes, cite each article independently as indicated on the article page online and as indicated below:

LastName, A.A.; LastName, B.B.; LastName, C.C. Article Title. *Journal Name* **Year**, *Volume Number*, Page Range.

ISBN 978-3-0365-3769-6 (Hbk)
ISBN 978-3-0365-3770-2 (PDF)

Cover image courtesy of American Hydro Corporation, York, Pennsylvania

Contents

About the Editors

John M. Cimbala (Professor)

John M. Cimbala is Professor of Mechanical Engineering at The Pennsylvania State University (Penn State). He received his B.S. degree in Aerospace Engineering (1979) from Penn State. Then, he obtained his M.S. degree (1980) and his Ph.D. degree (1984) in Aeronautics from The California Institute of Technology (Caltech). In 1984, Dr. Cimbala returned to Penn State as Assistant Professor of Mechanical Engineering. In 1990, he was promoted to Associate Professor and was granted tenure. In 1997, he was promoted to Professor.

During a sabbatical leave in 1993-94, Professor Cimbala worked at NASA Langley Research Center in Hampton, VA, where he advanced his knowledge of computational fluid dynamics (CFD) and turbulence modeling. During a sabbatical leave in 2002-03, he co-authored an undergraduate textbook, "Fluid Mechanics: Fundamentals and Applications", Y. A. Çengel and J. M. Cimbala, McGraw-Hill, New York (2006), now in its fourth edition (2018); it is used throughout the world and has been translated into eight languages. Professor Cimbala is author or co-author of several other textbooks and dozens of journal and conference papers. He is also the author of two novels, a devotional book, and a children's book. While on sabbatical leave during the academic year 2010-11, he worked at American Hydro Corporation in York, PA, where he used CFD to model flow through large hydroturbines and pump-turbines with the goal of optimizing their efficiency.

Dr. Cimbala conducts experimental and computational research in basic fluid mechanics, turbulence, and turbomachinery. He teaches courses in fluid mechanics; indoor air quality; instrumentation, measurements, and statistics; and air pollution. He has been an educational innovator throughout his career, such as using and promoting others to use the Internet, tablet PCs, and cell phone feedback in undergraduate and graduate courses. Awards include: College of Engineering Outstanding Teaching Award (1992), College of Engineering Premier Teaching Award (1996), George W. Atherton Award for Excellence in Teaching (1997), Teacher of the Year Award from Pi Tau Sigma (1997), and College of Engineering Outstanding Advising Award (1998). In 2009 he became a Fellow of the American Society of Mechanical Engineers.

Bryan J. Lewis (Professor)

Bryan J. Lewis is a Professor of Mechanical Engineering at Brigham Young University - Idaho, where he teaches courses in numerical analysis, fluid mechanics, thermodynamics, and product design. He also mentors students in undergraduate research and product development projects.

Prior to BYU-Idaho, Dr. Lewis worked in oil field research and product development for Halliburton Energy Services. As a senior research engineer, he assisted in development of hydraulic fracturing pumps, sand transport systems, and fluid-sand blending equipment. Dr. Lewis was the lead engineer for Halliburton's sound-mitigated hydraulic fracturing units. That project required strong partnership with internal and 3rd-party manufacturing firms, Halliburton field operations, oil companies, and state environmental regulators to develop, deploy, and field-validate the technology. Dr. Lewis published several articles on hydrajet-assisted hydraulic fracturing, as well on research in transient formation stress realignment during hydraulic fracturing events. Dr. Lewis holds more than 20 U.S. and international patents for various technologies he helped develop at Halliburton.

Dr. Lewis completed his PhD at The Pennsylvania State University, under the direction of Dr. John Cimbala. His PhD research focused on computational fluid analysis of hydroturbines and high-performance computing. He also was involved in the Global Engineering Teams program, working with a team of graduate student engineers from South Africa to develop a rural mobile health clinic for the South African Department of Health. The mobile clint designed by the team was manufactured and entered service with the health department in 2015.

Dr. Lewis also serves in a variety of roles for the American Society of Mechanical Engineers (ASME). He has been the faculty advisor for the BYU-Idaho ASME Student section since 2017, and has no plans of leaving that position any time soon. In 2019 Dr. Lewis served as the regional student section coordinator for the US/Canada Northwest Region. In 2020 he joined the executive committee for the newly formed Member Development and Engagement Sector. In 2021 he was elected as chair of the Idaho Section.

When he is not helping students, or working on improving his courses, Dr. Lewis enjoys spending time with his wife and their eight children. He and his wife are avid marathon runners, and enjoy books, movies, food, and traveling to new places with their children.

energies

Editorial

Advancements in Hydropower Design and Operation for Present and Future Electrical Demand

John Cimbala [1] and Bryan Lewis [2,*]

1 Department of Mechanical Engineering, The Pennsylvania State University, 234 Reber Building, University Park, PA 16802, USA; jmc6@psu.edu
2 Department of Mechanical and Civil Engineering, Brigham Young University-Idaho, 210 W 4th South, Rexburg, ID 83460, USA
* Correspondence: lewisbr@byui.edu

With the current infrastructure, meeting the ever-growing demand for electrical energy across the globe is becoming increasingly difficult. The widespread adoption of both commercial and residential non-dispatchable renewable energy facilities, such as solar and wind, further taxes the stability of the electrical grid, often causing traditional fossil fuel power plants to operate at lower efficiency, and with increased carbon emissions. Hydropower, as a proven renewable energy technology, has a significant part to play in the future global electrical power market, especially as increasing demand for electric vehicles will further amplify the need for dispatchable energy sources during peak charging times. Even with more than a century of proven experience, significant opportunities still exist to expand the worldwide hydropower resources and more efficiently utilize existing hydropower installations.

Given this context, this Special Issue of *Energies* was intended to present recent developments and advancements in hydropower design and operation. This Special Issue includes five articles, authored by international research teams from Japan, Pakistan, Sweden, Norway, the United States, and China. The authors bring the collective expertise of government research laboratories, university professors, industry research engineers, computer scientists, and economists. The articles explore advancements in hydroturbine and pump-turbine design, power plant operation, auxiliary equipment design to mitigate environmental damage, and an exploration of community-owned small hydropower facilities.

The articles contained in this Special Issue are:

- Flow Characteristics of Preliminary Shutdown and Startup Sequences for a Model Counter-Rotating Pump-Turbine [1].
- Flow Deflection between Guide Vanes in a Pump Turbine Operating in Pump Mode with a Slight Opening [2].
- Investigations of Rake and Rib Structures in Sand Traps to Prevent Sediment Transport in Hydropower Plants [3].
- Neural Network-Based Control for Hybrid PV and Ternary Pumped-Storage Hydro Plants [4].
- Community-Based Business on Small Hydropower (SHP) in Rural Japan: A Case Study on a Community Owned SHP Model of Ohito Agricultural Cooperative [5].

To conclude, the future of the global energy industry, while difficult to predict, will undoubtably include an ever-increasing portfolio of renewable energy sources. Hydroturbine and pump-turbine design engineers continue to find new improvements in efficiency [1–3]. As a result of ever-advancing system operations and control [4], hydropower will continue to be a major electricity contributor, given its historically proven reliability. The ability to dispatch energy generation when needed is a key element of maintaining stable electrical power grids to support growing global electrical demand. Many locations worldwide have significant untapped hydropower potential that will likely be developed in the coming

Citation: Cimbala, J.; Lewis, B. Advancements in Hydropower Design and Operation for Present and Future Electrical Demand. *Energies* **2022**, *15*, 2362. https://doi.org/10.3390/en15072362

Received: 11 March 2022
Accepted: 21 March 2022
Published: 24 March 2022

decades [5]. The successful future of worldwide electrical energy production and distribution will require collaboration among governments, environmental agencies, corporations, and researchers.

Author Contributions: Project administration, J.C.; Writing—original draft, B.L.; Writing—review and editing, J.C. All authors have read and agreed to the published version of the manuscript.

Funding: This research received no external funding.

Institutional Review Board Statement: Not applicable.

Informed Consent Statement: Not applicable.

Data Availability Statement: Not applicable.

Acknowledgments: The editors of this Special Issue are grateful to the MDPI Publisher for the invitation to act as Guest Editors of this Special Issue. All authors are indebted to the editorial staff of *Energies* for their kind cooperation, patience, and committed engagement.

Conflicts of Interest: The authors declare no conflict of interest.

References

1. Fahlbeck, J.; Nilsson, H.; Salehi, S. Flow Characteristics of Preliminary Shutdown and Startup Sequences for a Model Counter-Rotating Pump-Turbine. *Energies* **2021**, *14*, 3593. [CrossRef]
2. Ji, Q.; Wu, G.; Liao, W.; Fan, H. Flow Deflection between Guide Vanes in a Pump Turbine Operating in Pump Mode with a Slight Opening. *Energies* **2022**, *15*, 1548. [CrossRef]
3. Ivarson, M.M.; Trivedi, C.; Vereide, K. Investigations of Rake and Rib Structures in Sand Traps to Prevent Sediment Transport in Hydropower Plants. *Energies* **2021**, *14*, 3882. [CrossRef]
4. Nag, S.; Lee, K.Y. Neural Network-Based Control for Hybrid PV and Ternary Pumped-Storage Hydro Plants. *Energies* **2021**, *14*, 4397. [CrossRef]
5. Alam, Z.; Watanabe, Y.; Hanif, S.; Sato, T.; Fujimoto, T. Community-Based Business on Small Hydropower (SHP) in Rural Japan: A Case Study on a Community Owned SHP Model of Ohito Agricultural Cooperative. *Energies* **2021**, *14*, 3349. [CrossRef]

energies

Article

Flow Deflection between Guide Vanes in a Pump Turbine Operating in Pump Mode with a Slight Opening

Qingfeng Ji [1,2], Guoying Wu [3], Weili Liao [1] and Honggang Fan [2,*]

1 Institute of Water Resources and Hydro-Electric Engineering, Xi'an University of Technology, Xi'an 710048, China; tedwork123@outlook.com (Q.J.); liaoweili2004@163.com (W.L.)
2 Department of Energy and Power Engineering, Tsinghua University, Beijing 100084, China
3 China Water Resources Pearl River Planning, Surveying and Designing Co. Ltd., PRPSDC, Guangzhou 510610, China; wgy1@prpsdc.com
* Correspondence: fanhg@tsinghua.edu.cn; Tel.: +86-010-62794297

Abstract: During the startup and shutdown processes of a reversible-pump turbine (RPT) working in pump mode, abnormal sounds and vibrations usually occur in the distributor when the guide vanes (GVs) are at a slight opening (max opening of about 6%). The objective of this paper is to apply a three-dimensional numerical CFD method to study the unsteady flow behavior in the guide vane region of a pump turbine operating in pump mode. The dynamic meshing technique is introduced to simulate the startup and shutdown processes, and it is shown to be critical in accurately capturing the details of the flow pattern variations. In addition, the RNG k-epsilon two-equation turbulence model is applied and the governing equations are discretized with the finite volume method. Moreover, the boundary conditions are set through the calculation of the transient process of the power station. The results show that the main flow between the GVs is deflected during the startup and shutdown processes. In the shutdown process, the deflection occurs when the guide vane opening (GVO) is between 1.99 and 5.32 degrees, on average. In the startup process, the deflection occurs when the GVO is between 2.83 and 4.11 degrees, on average. In these processes, the velocity field and pressure field change dramatically. Simultaneously, the hydraulic torque (HT) on the GVs has a sharp change. The abrupt change in the HT leads to vibrations and abnormal sounds.

Keywords: pump turbine; pump mode; slight opening; flow deflection; dynamic meshing technique

Citation: Ji, Q.; Wu, G.; Liao, W.; Fan, H. Flow Deflection between Guide Vanes in a Pump Turbine Operating in Pump Mode with a Slight Opening. *Energies* **2022**, *15*, 1548. https://doi.org/10.3390/en15041548

Academic Editor: João Carlos de Campos Henriques

Received: 9 January 2022
Accepted: 16 February 2022
Published: 19 February 2022

1. Introduction

A reversible-pump turbine (RPT) is designed to pump water from a lower reservoir to a higher reservoir by using the surplus energy in a power grid. Furthermore, water has to go down to generate electrical energy through the RPT at peak hours or in case of an emergency. Hence, an RPT has to change working conditions between the pump mode and turbine mode [1]. In order to meet the requirements of a power grid, an RPT usually works in a low-load off-design working condition. Due to the RPT's special way of operating, it is hard to guarantee the stability of the units. Facing these challenges, many experts have made a lot of contributions. The unstable phenomena that occur in an RPT under runaway conditions (S-shaped characteristic in the turbine) are associated with fluctuations in the head and discharge of the system. This unstable behavior has been addressed by some authors [2–7] who believed that the nature of the vortex structures can be the mechanism that leads to potentially unstable characteristics.

Based on previous works, C. Gentner [8] performed a flow survey with CFD and PIV, and the results of the simulation and experiment had a great agreement, which strongly supported the earlier explanation of the mechanisms. In addition, some other authors pointed out that the reasons for the S-shaped characteristics are related to the development of rotating stall in the runner channels [9–12], or they are influenced by secondary flow in all parts of the RPT, and especially in the vaneless space [13]. Reference [14] performed a depth study of the onset and development of the unsteady phenomena that cause the unstable behavior of a pump turbine during a load-rejection scenario with servomotor failure, and their work revealed that, only when the unsteady phenomena evolved in a fully developed rotating stall that was characterized by a well-defined frequency, the unstable behavior of the RPT in turbine mode could be limited or eliminated with proper design criteria that aimed to move this critical operating point at lower flow rates.

Unstable phenomena such as rotating stall always come with the vibration of the units [5–7], which is harmful to the stability of the turbine and other devices. In addition, especially during the startup or shutdown process, the resonance of the units also frequently occurs, which is usually caused by rotor–stator interference (RSI) [15], self-excitation [16], or pressure pulsation [17]. To improve the stability and guarantee the successful startup of the units, misaligned guide vanes (MGVs) have been implemented and have achieved the expected results [18].

However, a few years ago, in both the Tianhuangping pumped-storage power station [19] and Yixing pumped-storage power station [20], abnormal sounds and vibrations occurred at the distributor of the RPT, which was working in the pump mode, and the guide vane opening was very small. In order to resolve this problem, B. Nennemann [21] conducted detailed research about the abnormal phenomena in the Yixing pumped-storage power station by using a 2D periodical CFD simulation. Consequently, the author claimed that unexpected bi-stable flow conditions and a self-excited torsion-mode flutter vibration were discovered in the GVs. Moreover, the vibration problem could be successfully eliminated by modifying the shape of the vanes. To address the issue of Tianhuangping, H. G. Fan [22] studied the HT of the GVs of the RPT during the startup and shutdown processes in the turbine mode with a 2D periodical CFD simulation. His work indicated that a repeated reversal of fluid occurred when the GVs worked with a slight opening during the shutdown process in the turbine mode. This phenomenon finally resulted in a dramatic increase in the HT, which caused the vibrations.

The aforementioned research on vibrations and abnormal sounds occurring at the distributer of an RPT was all performed with 2D periodical CFD simulations; however, the flow in an RPT is much more complicated when the working conditions are far from the optimal operation point and the aperiodicity of the flow is obvious [23]. Therefore, a 2D periodical CFD simulation is imperfect for dealing with this issue, and it cannot capture all of the details of the flow behavior. Based on previous studies, this paper describes the performance of a 3D CFD simulation of the startup and shutdown processes of the unit of the Tianhuangping pumped-storage power station when working in the pump mode. In the present study, the dynamic meshing technique was implemented, and more details of the flow behaviors were captured.

2. CFD Methodology

The RNG k-epsilon two-equation turbulence model has high precision and accuracy in solving high-Reynolds-number clearance flows. We found out that the deflection of the main flow can also be captured with the RNG k-epsilon model. Thus, the RNG k-epsilon two-equation turbulence model is adopted in this study. Due to the large amounts of calculations, we will perform a comparative study of other turbulence models in the future.

2.1. Geometry and Mesh

In this paper, the simulation was performed by using the commercial CFD solver ANSYS Fluent. The geometry of the prototype pump turbine of the Tianhuangping pumped-storage power station was implemented, and it was the same as that in previous work [23]. A geometric model is shown in Figure 1, and the characteristics of the parameters are listed in Table 1.

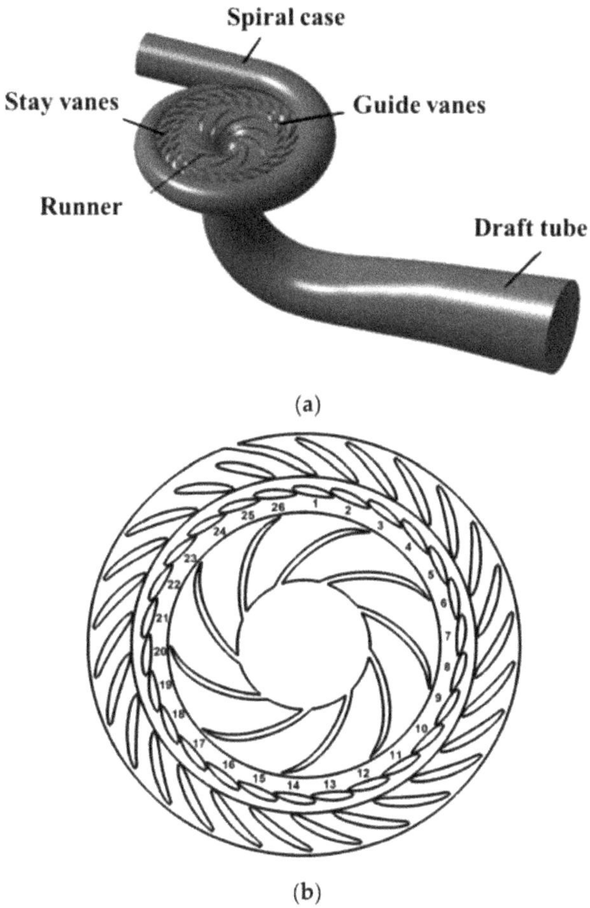

Figure 1. Pump turbine's (**a**) geometry and (**b**) GV number.

Table 1. Parameters of the pump turbine.

Parameters	Value
D_2 (mm)	2045
b_0 (mm)	262
Runner blade number	9
Guide vane number	26
Stay vane number	26
Rated speed (rpm) Head (m)	500,610

The dynamic meshing technique was used to control the movement of the GVs. Considering the geometry of the GVs and the deformation form of the mesh, a prism mesh was applied in the guide vane domain and stay vane domain (shown in Figure 2). In addition, in order to resolve the near-wall flow, a boundary layer mesh was implemented. In this case, the max Y+ was about 100 in the guide vane domain. A tetrahedral mesh was used in the rest of the domains due to the complex geometries.

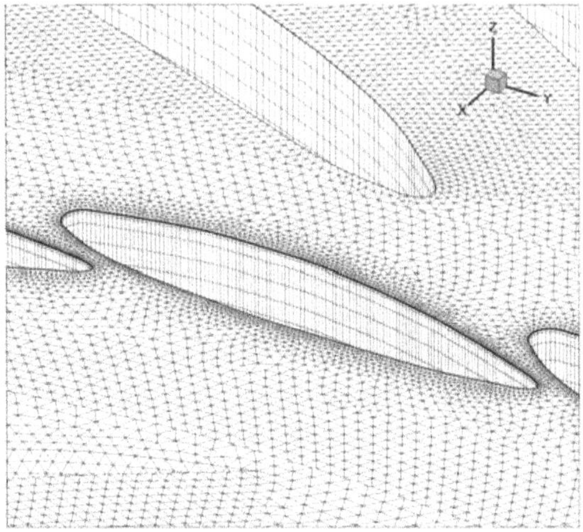

Figure 2. Mesh of the guide vane domain.

To ensure the accuracy of the results and to keep the computational cost at the minimum level, five sets of different meshes with 7.5 million elements, 8.9 million elements, 9.5 million elements, 11 million elements, and 12 million elements were used in a mesh independence study. The results will be discussed in the "Mesh and Time-Step Independence" section.

2.2. Boundary Conditions

The inlet discharge laws under the 610 m head, which were obtained from a transient process simulation of the Tianhuangping pumped-storage power station (shown in Figure 3), were selected for the boundary condition settings in this study. A velocity inlet and pressure outlet were implemented. A constant runner rotational speed was set at 500 rpm, while the opening rate of the GVs was 1/60 s and the closing rate was 1/25 s. In this case, $Re = 10^6$.

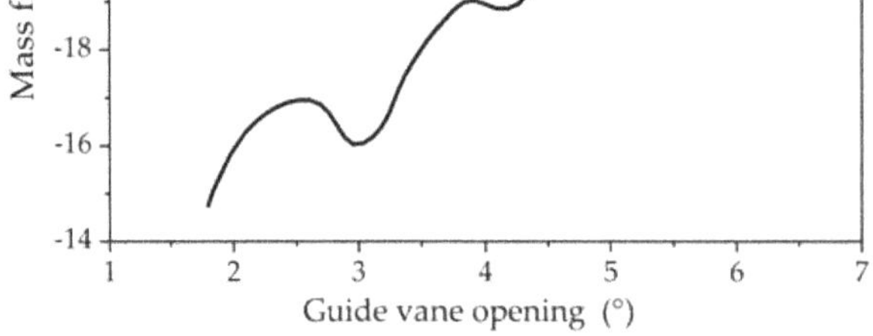

(a)

Figure 3. *Cont.*

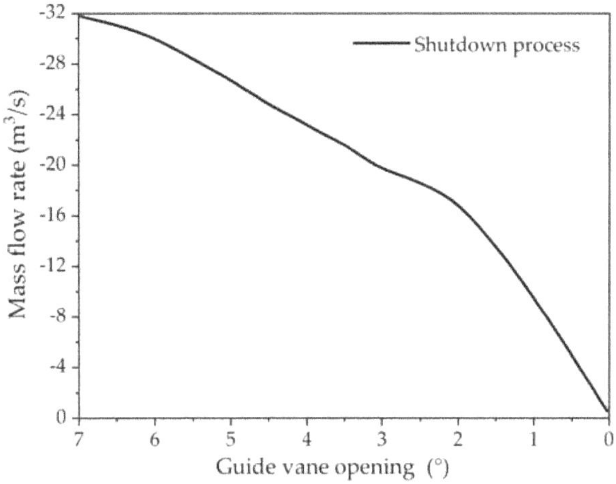

(**b**)

Figure 3. Discharge law of the transient process simulation of the Tianhuangping pumped-storage power station: (**a**) discharge law of the startup process; (**b**) discharge law of the shutdown process.

2.3. Parameter Nondimensionalization

In this paper, the Reynolds number and the dimensionless methods of the results are defined as follows:

$$Re = \frac{2\pi nRD_2}{\nu} \tag{1}$$

$$K_v = \frac{V}{\sqrt{2gH}} \tag{2}$$

$$C_p = \frac{\left(p - p_{ref}\right)}{\rho gH} \tag{3}$$

$$C_m = \frac{M}{0.5\rho {V_{ref}}^2 A_{ref} L_{ref}} \tag{4}$$

Here, Re is the Reynolds number, K_v is the velocity coefficient, C_p is the pressure coefficient, and C_m is the moment coefficient (a positive moment direction was defined as the closing direction of the guide vane). In addition, the rated rotational speed n = 500 rpm, outlet radius of the runner R = 1.0225 m, outlet diameter of the runner D_2 = 2.045 m, water density (20 °C) ρ = 998.2 kg/m^3, water viscosity (20 °C) μ = 1.003 × 10^{-3} Pa.s, head of the pump turbine H = 610 m, reference pressure P_{ref} = 0, reference velocity V_{ref} = 1 m/s, reference area $Aref$ = 1 m^2, and reference length L_{ref} = 1 m; in this paper, Re = 6.55 × 10^6.

2.4. Mesh and Time-Step Independence

Steady and unsteady calculations were performed with a constant guide vane opening (2°) with different mesh numbers and time steps. The number of nodes ranged from 7.5 million to 12 million, and the time step ranged from 1 × 10^{-4} to 5 × 10^{-3} s. The amplitude of the HT coefficient of the GVs was used as the criterion for this independence study. Figure 4a indicates that when the mesh number is greater than 11 million, the amplitude of the hydraulic torque only has a slight change; thus, 11 million mesh elements were adopted in this paper.

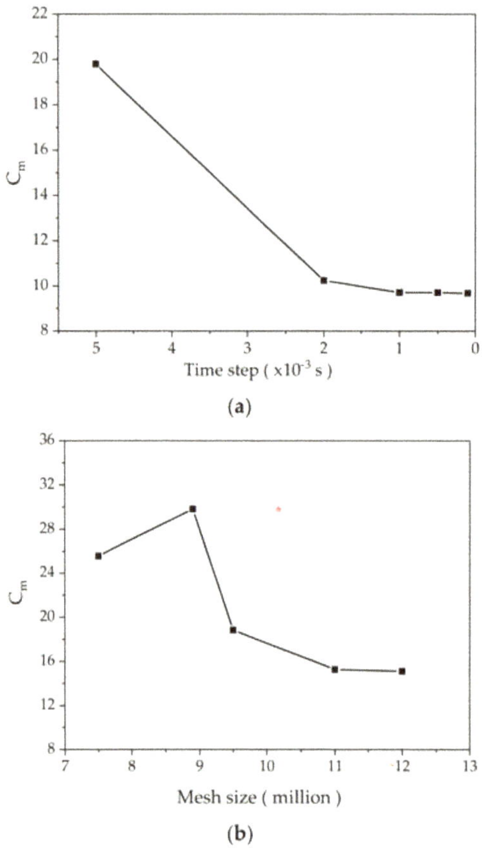

Figure 4. Independence study results: (**a**) the time step and (**b**) the mesh size.

A mesh with 11 million elements was used in the calculation of the time-step independence. Figure 4b shows that the amplitude of the hydraulic torque coefficient in the GVs is stable when the time step is below 5×10^{-4} s; thus, the time step was set to 5×10^{-4} s.

2.5. Numerical Model Validation

Mesh settings with openings of 3° and 6° were chosen for the numerical model validation. These meshes were both deformed from the initial mesh. The comparison between the results and the test data from the site shows that they are consistent with each other (the results for the 3° opening mesh are shown in Figure 5a; the results of the 6° opening mesh are shown in Figure 5b).

2.6. Calculation Approach

This work was carried out in the following steps, and Figure 6 shows a flowchart of our work:

(1) Steady simulation: In this step, the GVO was fixed at 2° and kept constant. The discharge law of the startup process was set as the inlet velocity boundary condition. The steady calculation was considered to be converged when the HT coefficient of the GVs converged, since the HT is the most important parameter in this case.

(2) Transient simulation: Before a dynamic process simulation, it is very necessary to carry out a transient simulation to develop the transient flow field. Thus, we carried out a transient simulation with a constant GVO, which was kept at 2°, and the rotational

speed (500 rpm) remained constant. The initialization was based on the result of step 1, and the boundary condition was kept the same as that in step 1. Finally, we set the transient simulation results of 10 complete runner revolutions as the initial flow field.

(3) Startup process simulation: The result of step 2 was used for the initialization of the startup process calculation with a dynamic meshing technique in the guide vane domain. The spring-smoothing method was selected, and the GVO increased with an opening rate of 1/60 s from 2° to 6.5°.

(4) Shutdown process simulation: The mesh of the last time interval of step 3 (GVO of 6.5°) was set as the initial mesh for the shutdown process calculation. The discharge law of the shutdown process was set as the inlet velocity boundary condition. In addition, a transient simulation with a constant GVO (6.5°) was carried out. Ten runner revolutions were completed with a constant rotational speed (500 rpm). The result of this simulation was used for the initialization of the shutdown process calculation. In the shutdown process calculation, the GVO decreased with a closing rate of 1/25 s from 6.5° to 1.5°.

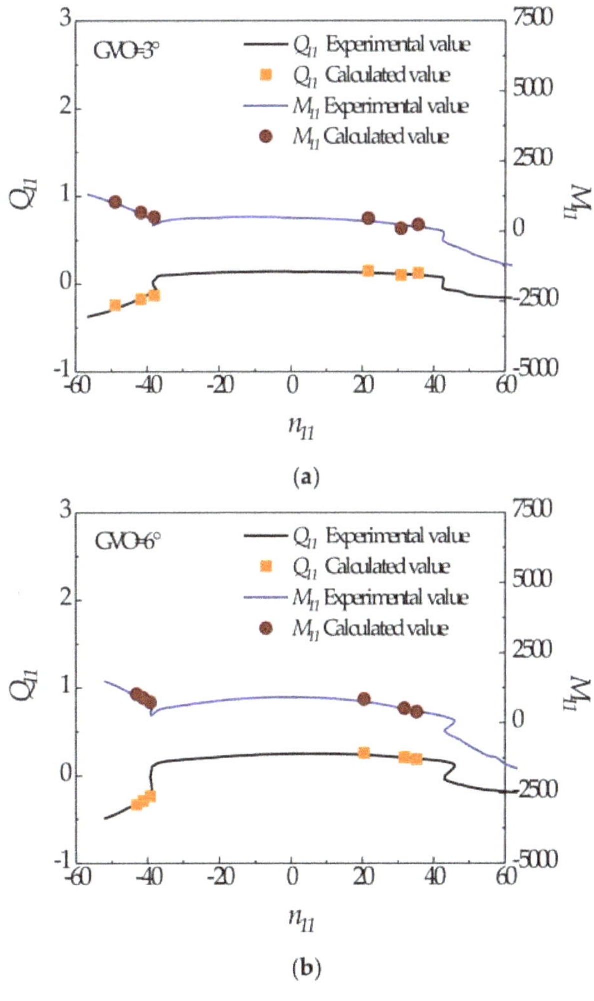

Figure 5. Numerical model validation results for different GVOs of (**a**) 3° and (**b**) 6°.

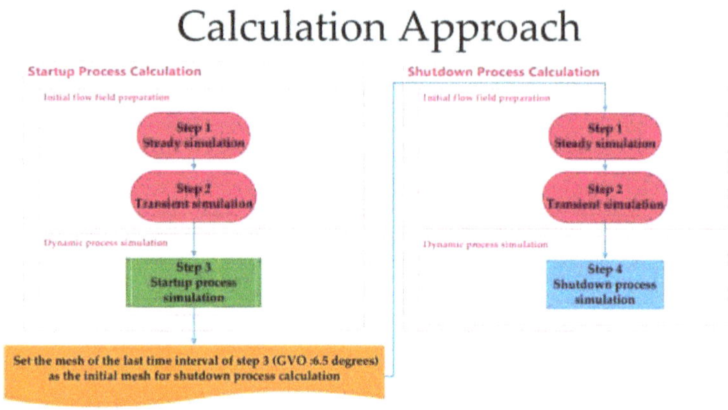

Figure 6. Workflow of the calculations.

3. Results and Discussion

3.1. Flow Deflection in the Guide Vane Domain

Figure 7 shows an example of a typical flow state in the shutdown process (GVO = 6.5°~1.82°), including the velocity contour and streamline at the Z = 0 plane (middle plane of the GV domain). It can be observed that the main stream in the GV domain is deflected when it passes through the gap in the GVs.

At GVO = 6.5°, the water from the runner outlet flows out along the channel between the movable guide vanes, and its direction is consistent with the overflow channel. When the GVO decreases to 3.1°, it can be seen that the direction of the water flow is no longer towards the right side of GV #4, but is deflected by nearly 90° and flows towards the fixed guide vane. The main flow in the deflection process is very unstable. When the GVO is further reduced to 1.82°, the water flow is completely attached to the right wall of GV #5. According to the streamline distribution, a clockwise circulation is formed between the movable guide vane and the fixed guide vane.

The reverse process is found during the startup, as shown in Figure 8. At GVO = 2°, the water flow from the runner outlet is completely attached to the right wall of GV #5. When the GVO increases to 3.3°, it can be seen that the direction of the water flow is no longer attached to GV #5, but is deflected by nearly 90°, and the water flows towards the fixed guide vane. There are vortices with opposite rotation directions on both sides of the main stream. When the GVO is further decreased to 4.45°, the main stream flows out along the channel between the movable guide vanes, and its direction is consistent with the overflow channel.

We gave definitions for the two flow types in the pump mode, which are sketched in Figure 9. Type I is defined as the flow type with a direction consistent with the incoming flow direction, while Type II is the flow type with a deflected direction that is attached to the guide vane head. This belongs to the state of the transition between the two.

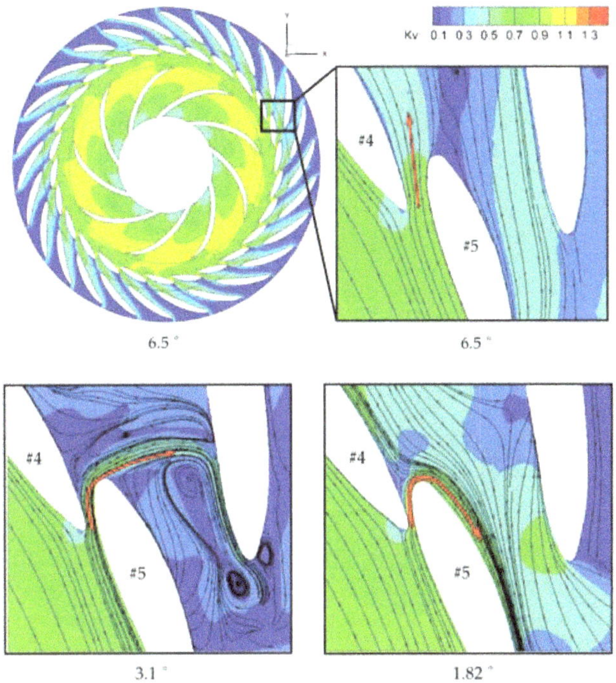

Figure 7. Global velocity contours of the shutdown process (Z = 0 plane).

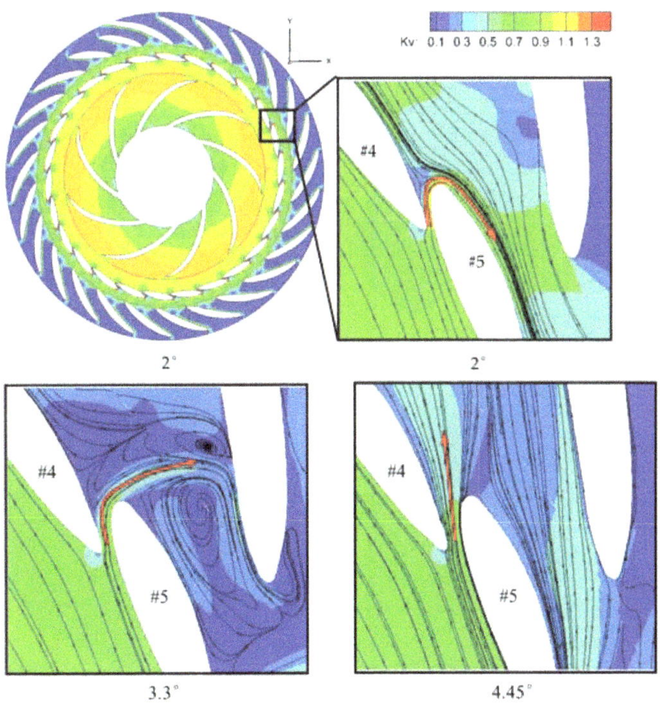

Figure 8. Global velocity contours of the startup process (Z = 0 plane).

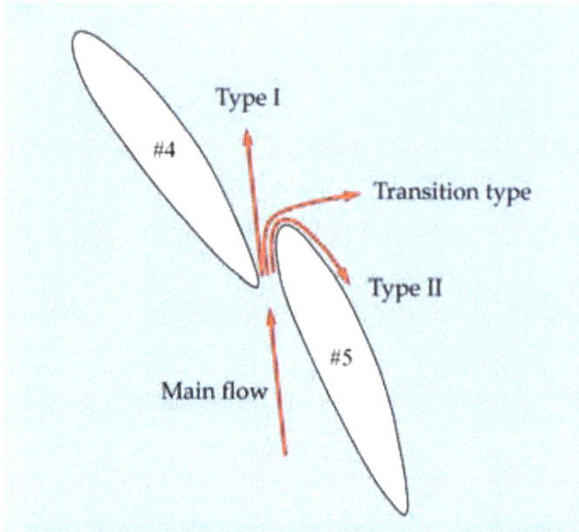

Figure 9. A schematic diagram of the flow pattern definitions.

The main flow in the gap of GVs #26 and #1 was chosen to demonstrate the formation of these two flow patterns. Figure 10 shows the velocity resolution of the main flow near the tailing edge of GV #26. V_r is the radial velocity, and V_θ is the tangential velocity. In the plot, a subscript of 1 stands for a small opening, and 2 stands for a large opening. It can be seen that, near the tailing edge of the GV, the radial velocity is greater than the tangential velocity. At the small opening, the incidence angle is too large, and this large incidence angle can lead to flow separation near the tailing edge of GV #26. Then, this separation forces the main flow to turn toward the leading edge of GV #1. Because the curvature radius of the GV surface is too large, the Coanda effect [24,25] can make the main flow adhere to the surface of GV #1 so that the main flow stays in Type II. However, at the large opening, the incidence angle can be reduced with a change in the vane angle; thus, the flow separation is weakened and the main flow eventually stays in Type I.

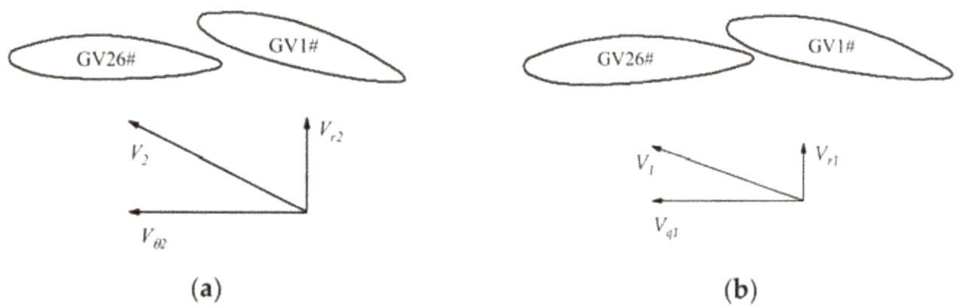

Figure 10. Analysis of the resultant velocity: (**a**) small opening; (**b**) large opening.

3.2. *Flow-Field Characteristics of Startup and Shutdown*

Because the flow pattern of each guide vane has little difference, we selected some guide vanes (GV03, GV04, GV05, GV06) for the flow-field analysis. In order to analyze the characteristics of flow deflection between moving and stationary guide vanes in detail, Figure 11 shows the local velocity contours and pressure contours of the shutdown process. During this process, the GVs gradually closed from a max opening of 19.69% to a max opening of 5.91% (decrease in the GVO from 6.5° to 1.95°).

Figure 11. *Cont.*

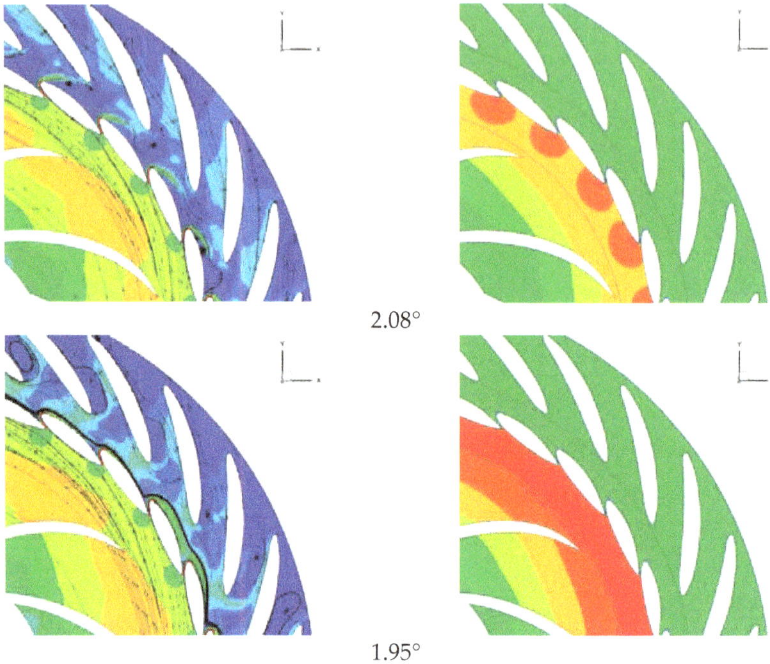

Figure 11. Local velocity contours (**left**) and pressure contours (**right**) of the shutdown process (Z = 0 plane, GVO = 6.5°~1.95°).

When GVO = 6.5°~4.58°, the main flow stays in Type I. At GVO = 6.5°, the direction of the main flow is towards the left side of the fixed guide vane directly in front of the flow channel. The velocity coefficient k_v is about 0.9. The velocity at the throat between the two GVs is the highest where k_v is about 1.2. Therefore, the pressure here is also the lowest relative to the flow field inside and outside the GV. The velocity on the right side of the main stream is very low with $k_v \approx 0.1$.

As the GVO decreases, the main flow tends to shift away from the left movable guide vane, as shown at GVO = 4.58°. When the GVO decreases, the velocity at the throat gradually increases and the pressure further decreases, which is caused by the decrease in the cross-sectional area of the overflow.

In the range of GVO = 4.58°~2.08°, the main flow is in the transition stage between Type I and Type II. When GVO = 3°, the direction of the main flow is deflected by nearly 90 degrees. At the same time, there is a high-pressure area at the left GV, where $c_p \approx 0.8$. The trailing-edge of the main stream impinges on the other fixed guide vane (transferred from left to right) and flows along the fixed guide vane wall to both sides. With the decrease in the GVO, the deflection angle of the main flow increases further and changes to flow in the flow channel between the fixed guide vanes, as shown at GVO = 2.08°. At this time, the pressure coefficient inside the GV increases to $c_p \approx 1.2$.

When the GVO further decreases to 1.95°, the main flow state changes to Type II. At this opening, the main stream flows around the leading edge of the GV. The throat position becomes the boundary between the high-pressure area and low-pressure area.

Moreover, the main flow stays in Type I when the GVO is more than 6.24°, and the flow pattern stays in Type II when the GVO is less than 1.95°. The pressure plot shows that the pressure at the tailing edge of the GV is higher than that at the leading edge of the GV; thus, the water coming from the runner domain pushes the GV to close, and the resultant moment should be positive. Furthermore, with the decrease in the GVO, the HT must theoretically increase.

During startup, the flow goes through the process of opposite deflection, as shown in Figure 12. With the increase in the GVO, the main flow starts deflecting from Type II to Type I. Unlike in the shutdown process, when GVO = 2.8°, the main flow is still in Type II, while the main flow is converted to Type I.

Figure 12. *Cont.*

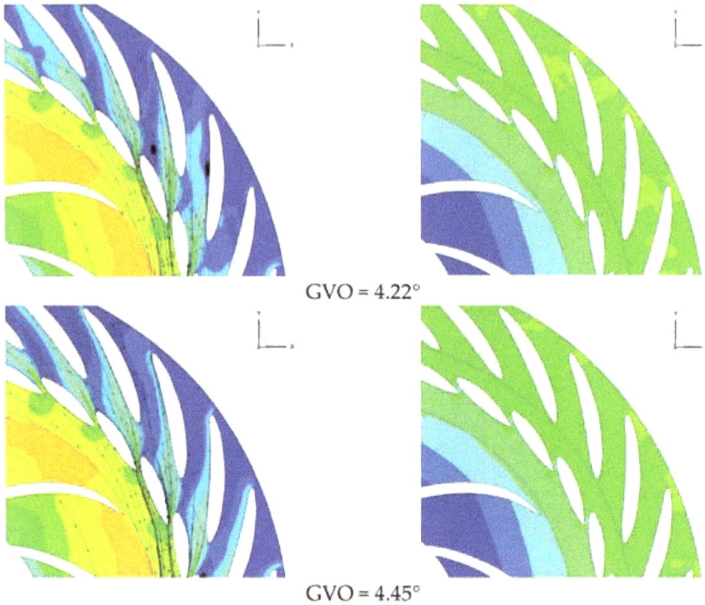

Figure 12. Local velocity contours (**left**) and pressure contours (**right**) of the startup process (Z = 0 plane).

The pressure plot shows that the pressure at the tailing edge of the GV is larger than the pressure at the leading edge of the GV; thus, the resultant moment on the GV helps the GV to close. In this paper, the moment that forces the GV to close is defined as a positive moment (HT > 0).

The behaviors of the deflections during the startup and shutdown processes are not the same. Figure 13 shows a comparison of the GVO ranges of the deflections between the shutdown and startup processes. The specific deflecting range of the shutdown process is larger than that of the startup process. For the shutdown process, the deflection angle ranges from 1.99° to 5.32° on average. For the startup process, the range is reduced to 2.83° to 4.11° on average.

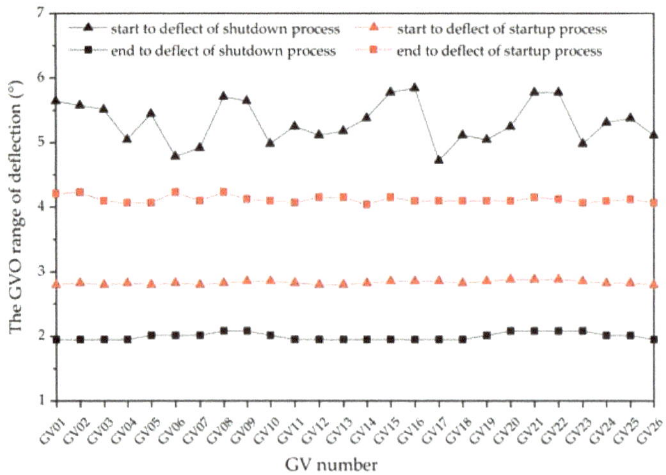

Figure 13. A comparison of GVO ranges of the deflections between the shutdown and startup processes.

During the shutdown process, the GVO of the starting deflection fluctuates greatly on different guide vanes. However, for the ending deflection, the fluctuation is small.

3.3. Hydraulic Torque on the Guide Vanes

Several GVs were chosen for analysis in this part because the torque change trends of most GVs are similar.

Figure 14 presents the GVO-C_m plot for GV03, GV04, GV05, GV06, and GV10 during the simulation of the startup process. In the beginning, C_m gradually decreases with the increase in the GVO, but when the GVO goes up to 2.8°, C_m abruptly increases. This sudden increase in C_m is indicated by the red line. When the GVO is about 4.05°, which is indicated by the blue line, C_m of GV10 is larger than that of the others. As can be seen in Figure 15, when GVO = 4.05°, the deflection speed of the main stream of GV10 is faster than that of GV06, so the torque of GV10 at this time is greater than that of the other GVs. When the GVO is around 4.22° (indicated by the green line), the flow deflection is complete, and the trend of C_m slows down. In addition, it can be seen that during the deflection process (GVO ranging from 2.8° to 4.2°), the values of C_m have phase differences between each other. This difference is caused by the unstable flow behavior of the main flow during the deflection process, as mentioned above.

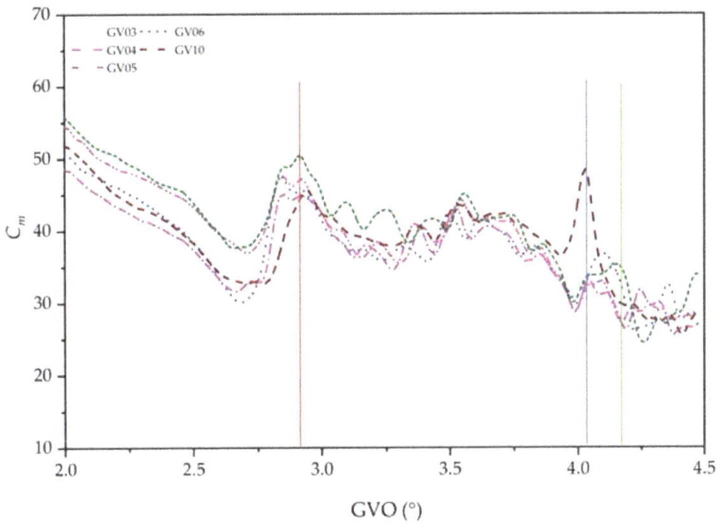

Figure 14. GVO-C_m plot of the startup process.

By comparing the C_m plot with the velocity contours of the main flow (Figure 12), it can be seen that the main flow deflection and the increase in C_m occur at the same opening. When the deflection occurs abruptly, the velocity near the GV surface on the side of the stay vane decreases quickly; thus, the pressure on the surface of the GV increases. In addition, the resultant moment on the GV suddenly increases.

The GVO-C_m plot for GV01, GV03, GV04, GV05, and GV06 obtained from the calculation of the shutdown process shows that C_m increases with the gradual decrease in the GVO. However, when the GVO ranges from 5.26° to 3.15° (which is indicated by the red line and blue line in Figure 16), C_m has a turbulent trend, and the GV near the nose of the spiral case (GV01) has a large moment value. This phenomenon is due to the instability of the main flow when the GVO is in the specific range (from a max opening of 19.39% (about 6.4°) to a max opening of 5.91% (about 1.95°)) that was mentioned before. The velocity near the surface of the GV fluctuates, and this fluctuation leads to the turbulent fluctuation of C_m. When the GVO is less than 3.15°, C_m has a stable trend. Moreover, when the GVO is about 2.2°, C_m descends sharply, as indicated by the green line in the plot. In addition, the main flow stays in Type II when C_m flattens (which is indicated by the yellow line in Figure 16).

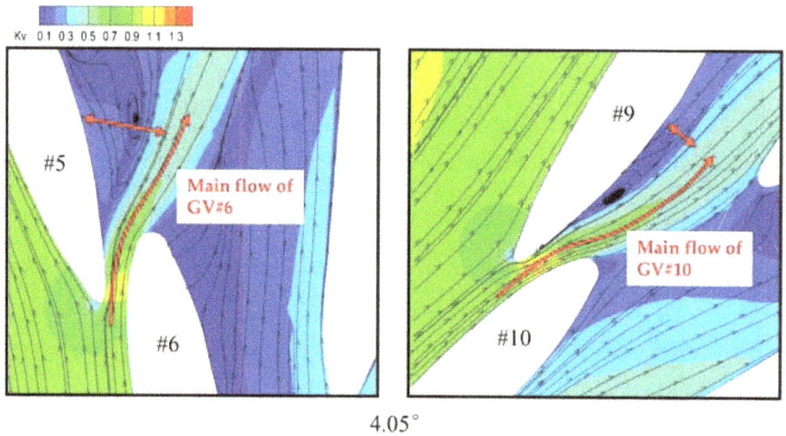

Figure 15. Comparison of deflection speeds between the main flow of GV10 (**right**) and that of GV06 (**left**) at GVO = 4.05°.

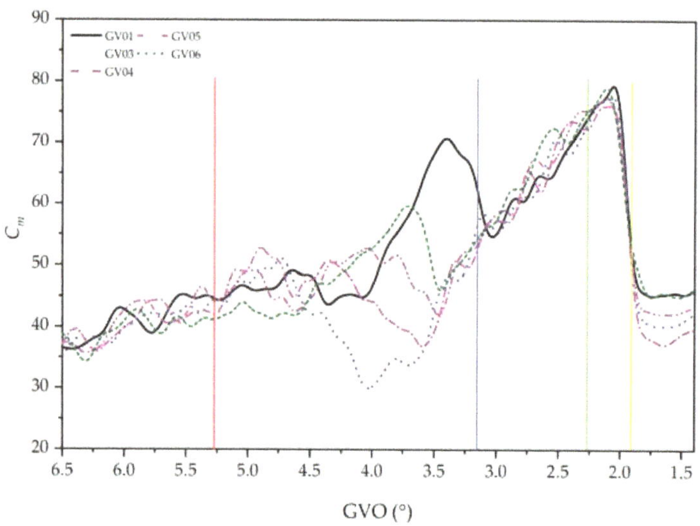

Figure 16. GVO-C_m plot of the shutdown process.

It can be inferred that the deflection led to a sudden change in the velocity near the GV, and the pressure on the GV's surface abruptly changed. Consequently, the change in pressure resulted in a sudden increase in the C_m; if the transmission of the distributer cannot immediately adapt to this change, there will be a relative displacement between the shaft of the GV and the friction device, which will produce violent friction. This violent friction will lead to vibrations and abnormal sounds.

4. Conclusions

Numerical studies were presented for the startup and shutdown processes of an RPT in the pump mode when GVO is between 1.5 and 6.5 degrees. A dynamic meshing technique was used to investigate the flow deflection of the main flow between the guide vanes. The primary findings include:

1. During the startup and shutdown processes of the RPT, flow separation occurred at the tailing edge of the GV when the GVO was less than the 6% max GVO. This separation forced the main flow to turn to the other side, and then the main flow adhered to the surface of the GV because of the Coanda effect; thus, the main flow stayed in Type II. With the increase in the GVO, the separation was weakened, so the main flow stayed in Type I. Therefore, the main flow was deflected from Type II to Type I during the startup process. During the shutdown process, the main flow was deflected from Type I to Type II.
2. When the main flow was deflected away from the surface of the GV or adhered to the surface of GV, the velocity field had an intense change near the GV. In addition, the pressure field also changed. Consequently, the hydraulic torque sharply changed at this moment. If the components of a distributor cannot adapt to this dramatic change, the violent friction will lead to vibrations and abnormal sounds.
3. During the startup and shutdown processes, the deflection of the main flow happened in a specific GVO range. The differences in the discharge laws of these two dynamic processes cause the shutdown process to have a wide GVO range for deflection.
4. Moreover, the position of the GV also has an influence on the deflection and the hydraulic torque. The deflection of the main flow at different GVs did not occur at the same GVO, and the hydraulic torque on the GV near the nose of the spiral case had the maximum hydraulic torque magnitude.

Author Contributions: Methodology, Q.J., G.W.; formal analysis, Q.J., H.F., W.L.; investigation, G.W.; resources, H.F.; data curation, Q.J., G.W.; writing—original draft preparation, Q.J.; writing—review and editing, Q.J., H.F., W.L.; supervision, H.F., W.L.; All authors have read and agreed to the published version of the manuscript.

Funding: This research was funded by [National Natural Science Foundation of China] [Grant No. 51879140, No. 51679196], [State Key Laboratory of Hydroscience and Hydraulic Engineering] [Grant No. 2021-KY-04], [Creative Seed Fund of Shanxi Research Institute for Clean Energy of Tsinghua University], and [Tsinghua-Foshan Innovation Special Fund(TFISF)] [Grant No. 2021THFS0209].

Acknowledgments: The authors would like to thank the National Natural Science Foundation of China (No. 51879140, No. 51679196), State Key Laboratory of Hydroscience and Hydraulic Engineering (Grant No. 2021-KY-04), Creative Seed Fund of Shanxi Research Institute for Clean Energy of Tsinghua University and Tsinghua-Foshan Innovation Special Fund(TFISF) 2021THFS0209 for their financial support.

Conflicts of Interest: The authors declare no conflict of interest.

Nomenclature:

ρ	density
V	absolute velocity
V_r	radial velocity
V_θ	tangential velocity
ε	turbulent energy dissipation rate
S_{ij}	Strain-rate tensor
μ_{eff}	equivalent viscosity coefficient
Z	axis Z
HT	hydraulic torque

References

1. Eduard, E.; Carme, V.; David, V.; Alexandre, P.; Cristian, G.R. *Condition Monitoring of Pump-Turbines New Challenges*; Elsevier: Amsterdam, The Netherlands, 2015; pp. 151–163.

2. Widmer, C.; Staubli, T.; Ledergerber, N. Unstable Characteristics and Rotating Stall in Turbine Brake Operation of Pump-Turbines. *J. Fluids Eng.* **2011**, *133*, 041101. [CrossRef]
3. Gentner, C.; Sallaberger, M.; Widmer, C.; Braun, O.; Staubli, T. Numerical and experimental analysis of instability phenomena in pump turbines. In Proceedings of the 26th IAHR Symposium on Hydraulic Machinery and Systems, Beijing, China, 19–23 August 2012.
4. Gong, R.Z.; Li, D.Y.; Wang, H.J.; Wei, X.Z.; Qin, D.Q. Investigation into the flow details of runner region in a pump turbine at off-design conditions. *SAGA Adv. Mech. Eng.* **2016**, *8*, 1–15. [CrossRef]
5. Martin, C.S. Representation of Pump Characteristics for Transient Analysis. In Proceedings of the ASME, Symposium on Performance Characteristics of Hydraulic Turbines and Pumps, Winter Annual Meeting, Boston, MA, USA, 13–18 November 1983; pp. 1–13.
6. Triki, A.; Essaidi, B. Investigation of Pump Failure-Induced Waterhammer Waves: A Case Study. *J. Press. Vessel. Technol. Trans. ASME* **2022**, *144*, 061403. [CrossRef]
7. Triki, A.; Essaidi, B. On the transient flow behavior in pressurized plastic pipe—Based water supply systems. *AQUA Water Infrastruct. Ecosyst. Soc.* **2021**, *70*, 67–76. [CrossRef]
8. Gentner, C.; Sallaberger, M.; Widmer, C.; Bobach, B.-J.; Jaberg, H.; Schiffer, J.; Senn, F.; Guggenberger, M. Comprehensive experimental and numerical analysis of instability phenomena in pump turbines. In Proceedings of the 27th IAHR Symposium on Hydraulic Machinery and Systems, Montreal, Canada, 22–26 September 2014.
9. Hasmatuchi, V.; Farhat, M.; Roth, S.; Botero, F.; Avellan, F. Experimental Evidence of Rotating Stall in a Pump-Turbine at Off-Design Conditions in Generating Mode. *ASME J. Fluid. Eng.* **2011**, *133*, 051104. [CrossRef]
10. Botero, F.; Hasmatuchi, V.; Roth, S.; Farhat, M. Non-intrusive detection of rotating stall in pump-turbines. *Elsevier Mech. Syst. Signal Processing* **2014**, *48*, 162–173. [CrossRef]
11. Hasmatuchi, V.; Roth, S.; Botero, F.; Avellan, F.; Farhat, M. High-speed flow visualization in a pump-turbine under off-design operating conditions. In Proceedings of the 25th IAHR Symposium on Hydraulic Machinery and Systems, Timisoara, Romania, 20–24 September 2010.
12. Yang, J.; Pavesi, G.; Yuan, S.; Cavazzini, G.; Ardizzon, G. Experimental Characterization of a Pump–Turbine in Pump Mode at Hump Instability Region. *ASME J. Fluid. Eng.* **2015**, *137*, 051109. [CrossRef]
13. Guggenberger, M.; Senn, F.; Schiffer, J.; Jaberg, H.; Gentner, C.; Sallaberger, M.; Widmer, C. Experimental investigation of the turbine instability of a pump-turbine during synchronization. In Proceedings of the 27th IAHR Symposium on Hydraulic Machinery and Systems, Montreal, QC, Canada, 22–26 September 2014.
14. Cavazzini, G.; Covi, A.; Pavesi, G.; Ardizzon, G. Analysis of the Unstable Behavior of a Pump-Turbine in Turbine Mode: Fluid-Dynamical and Spectral Characterization of the S-shape Characteristic. *ASME J. Fluid. Eng.* **2015**, *138*, 021105. [CrossRef]
15. He, L.Y.; Wang, Z.W.; Kurosawa, S.; Nakahara, Y. Resonance investigation of pump-turbine during startup process. In Proceedings of the 27th IAHR Symposium on Hydraulic Machinery and Systems, Montreal, QC, Canada, 22–26 September 2014.
16. Nennemann, B.; Sallaberger, M.; Henggeler, U.; Gentner, C.; Parkinson, E. Assessment of guide vane self-excitation stability at small openings in pump flow. In Proceedings of the 26th IAHR Symposium on Hydraulic Machinery and Systems, Beijing, China, 19–23 August 2012.
17. Sun, H.; Xiao, R.F.; Liu, W.C.; Wang, F.J. Analysis of S Characteristics and Pressure Pulsations in a Pump-Turbine with Misaligned Guide Vanes. *ASME J. Fluid. Eng.* **2013**, *135*, 051101. [CrossRef] [PubMed]
18. Xiao, Y.X.; Xiao, R.F. Transient simulation of a pump-turbine with misaligned guide vanes during turbine model start-up. *Acta Mech. Sin.* **2014**, *30*, 646–655. [CrossRef]
19. Kong, L.H. Analysis of abnormal sounds in working condition change-over for high-head pump-turbine. *Mech. Electr. Tech. Hydro Power Stn.* **2004**, *27*, 12–14.
20. Cai, J.; Zhou, X.J.; Den, L.; Zhang, W.H. The Research of the Abnormal Water Hammer Phenomenon based on the Unit 3 over Speed Test of Jiangsu Yixing Pumped Storage Power Station. *China Acad. J. Electron. Publ. House Water Power* **2009**, *35*, 76–79.
21. Nennemann, B.; Parkinson, É. YiXing pump turbine guide vane vibrations: Problem resolution with advanced CFD analysis. In Proceedings of the 25th IAHR Symposium on Hydraulic Machinery and Systems, Timisoara, Romania, 20–24 September 2010.
22. Fan, H.G.; Yang, H.X.; Li, F.C.; Chen, N.X. Hydraulic torque on the guide vane within the slight opening of pump turbine in turbine operating mode. In Proceedings of the 27th IAHR Symposium on Hydraulic Machinery and Systems, Montreal, QC, Canada, 22–26 September 2014.
23. Fan, H.G.; Ji, Q.F.; Liao, W.L.; Yang, H.X. Flow Analysis of the Guide Vanes Region of Pump Turbine at the Slight Opening in the Pumping Startup Process. In Proceedings of the ASME 2016 Fluids Engineering Division Summer Meeting collocated with the ASME 2016 Heat Transfer Summer Conference and the ASME 2016 14th International Conference on Nanochannels, Microchannels, and Minichannels, Washington DC, USA, 10–14 July 2016.
24. Ferziger, J.H.; Peric, M. *Computational Methods for Fluid Dynamic*, 3rd ed.; Springer: Berlin/Heidelberg, Germany, 2002.
25. Moshfeghi, M.; Hur, N. Numerical Investigation on the Coanda Effect over the S809 Airfoil with Synthetic Jet Actuator at High Angle of Attack. In Proceedings of the ASME 2014 4th Joint US-European Fluids Engineering Division Summer Meeting collocated with the ASME 2014 12th International Conference on Nanochannels, Microchannels, and Minichannels, Chicago, IL, USA, 3–7 August 2014.

energies

Article

Neural Network-Based Control for Hybrid PV and Ternary Pumped-Storage Hydro Plants

Soumyadeep Nag * and Kwang Y. Lee

School of Engineering and Computer science, Baylor University, Waco, TX 76706, USA; Kwang_Y_Lee@baylor.edu
* Correspondence: soumyadeep_nag@baylor.edu

Abstract: The growth in renewable energy integration over the past few years, primarily fueled by the drop in capital cost, has revealed the requirement for more sustainable methods of integration. This paper presents a collocated hybrid plant consisting of solar photovoltaic (PV) and Ternary pumped-storage hydro (TPSH) and designs controls that integrate the PV plant such that the behavior and the controllability of the hybrid plant are similar to those of a conventional plant within operational constraints. The PV array control and hybrid plant control implement a neural–network-based framework to coordinate the response, de-loading, and curtailment of multiple arrays with the response of the TPSH. With the help of the designed controls, a symbiotic relationship is developed between the two energy resources, where the PV compensates for the TPSH nonlinearities and provides required speed of response, while the TPSH firms the PV system and allows the PV to be integrated using its existing infrastructure. Simulations demonstrate that the designed controls enable the PV system to track references, while the TPSH's firming and shifting transforms the PV system into a base load plant for most of the day and extends its hours of operation.

Keywords: hybrid power; neural networks; pumped-storage hydro; solar; photovoltaic; hydropower; renewable energy

Citation: Nag, S.; Lee, K.Y. Neural Network-Based Control for Hybrid PV and Ternary Pumped-Storage Hydro Plants. *Energies* **2021**, *14*, 4397. https://doi.org/10.3390/en14154397

Academic Editors: John M. Cimbala and Bryan J. Lewis

Received: 25 June 2021
Accepted: 15 July 2021
Published: 21 July 2021

1. Introduction

1.1. Problem Statement

With the increasing addition of renewables, the requirements for flexibility in power networks have increased. During periods of high renewable penetration, flexible resources, such as fossil-fueled plants, are not available. This reduces system inertia and reserves due to which system flexibility and resilience are compromised. Furthermore, the frequency of high-impact events such as storms, forest fires, and cyberattacks has increased in recent years [1]. To cope with these disturbances, flexible resources are required. Although battery energy storage is a solution, the scale of battery storage required for utility-scale PV or wind projects presents a technoeconomic barrier [2].

1.2. Proposed Solution

To remedy the issue of higher flexibility requirements of modern and future power networks, more sustainable methods of renewable integration are required. This paper proposes a hybrid PV and ternary PSH (TPSH) combination and designs controls for the same to ensure that the PV system and the TPSH system work in a coordinated manner to cater to the network requirements and to suppress internal disturbances. The proposed system controls can be decomposed into three parts: (a) hybrid plant control, (b) PV plant control, and (c) TPSH plant control. These are detailed in this paper.

Resilience is enhanced by fast responses of the PV converter systems and flexibility of the pump mode with hydraulic short-circuit (HSC) of the TPSH. Reliability is provided through the firming capability of TPSH in pump mode using the HSC. Using the HSC, the TPSH firms the irradiance-based PV disturbances within operational constraints.

When used in a collocated manner with the PV plant, the TPSH aids in firming and shifting PV generation. To firm PV generation is to make constant the power generated from the PV plant by absorbing the excess and supplying the deficit. Shifting PV generation is to absorb the excess and supply the absorbed during peak hours. The key strategy is to utilize the HSC of the TPSH to firm the PV system during the entire day while also storing energy. This stored energy is then used during the peak hours of the day using generation mode of TPSH. This provides an option for avoiding curtailment. Since both PV and TPSH are collocated, the transmission losses while pumping are negligible. The controls for the PV plant enable it to compensate for the nonlinearities of the hydro plant and enhance its response.

1.3. Motivation and Literature Review

The modern power system is a hybrid energy system and has the capability to manage multiple energy sources. Thus, the concept of hybridization is not new; however, a collocated PV and TPSH plant is proposed for the first time. No single energy source is optimal for all situations. Hybridization and coordinated control are required to utilize the benefits of each source while overcoming their shortcomings [3]. The designed controls facilitate this using the fast response of PV inverters and the storage capacity of PSH. A TPSH is considered to leverage the pump mode power consumption flexibility of the TPSH to firm the PV system.

To enhance the flexibility of the pumped hydro plant, many configurations have been proposed and implemented, such as the adjustable-speed PSH [4,5] and the TPSH [6,7]. A TPSH is composed of a synchronous machine, turbine, and pump on the same shaft as shown in Figure 1. The pump and turbine are separated with a clutch. Key advantages of the TPSH include (a) rapid mode change without the loss of synchronism, (b) pump mode power consumption flexibility using HSC, and (c) smooth start capability in pump mode [6,7]. In the turbine mode and pure pump mode, its working principles are like hydro turbines and pumps, respectively. When the HSC (or the shared penstock connecting the pump and turbine) is activated in pump mode, a part of (or all of) the water pumped is transferred to the turbine. This in turn causes the turbine to produce torque that is supplied to the shaft. This mechanical torque from the turbine will reduce the current drawn from the grid while still rotating at the synchronous speed. The remaining water is pumped up to the upper reservoir. Another perspective is that, when rotating at the rated speed, the pump still generates the rated head, but the amount of water transferred to the upper reservoir or true work done decreases.

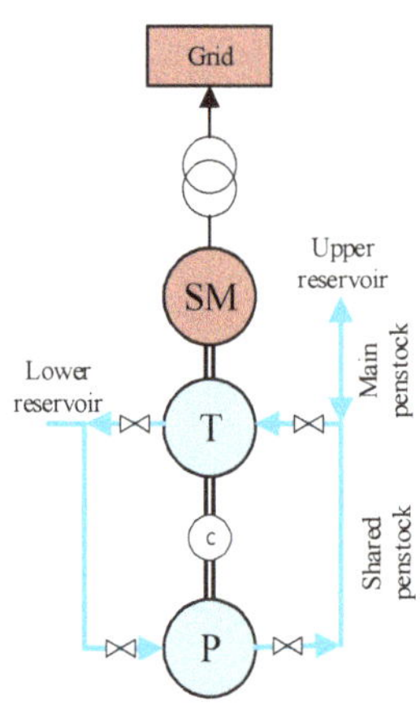

Figure 1. Schematic diagram of the TPSH, where P = centrifugal pump, T = Francis turbine, c = clutch, and SM = synchronous machine.

To magnify the benefits of the proposed PV + TPSH hybrid, it can be compared with the quaternary PSH (QPSH) [8]. In QPSH, the pump and turbine have separate shafts and electrical machines, although the HSC still exists. The proposed hybrid provides a better alternative to the QPSH due to several reasons. Firstly, in QPSH, although the converter-based pump adds flexibility, it does not add generation capacity to the existing PSH system. The proposed PV + TPSH system has added generation capacity and added flexibility through the PV system. In fact, the QPSH reduces generation capacity if a retrofitting case is considered, as the space for one unit will have to be spared for the variable speed pump. Secondly, in generation mode the QPSH is similar in performance to any conventional PSH; however, the proposed PV + TPSH takes advantage of the PV inverters to provide rapid response to system disturbances in generation and pump modes. Lastly, the QPSH is a significant investment burden as not only do the pump and turbine have separate shafts but the converter also adds to the cost of the system without adding generation capacity.

The proposed PV + TPSH takes advantage of the existing infrastructure which is unused for PSH plants during the day. In addition, a report from the World bank [9] highlighted that floating PV systems can have several advantages as follows:

- Reduced capital cost with utilization of existing transmission infrastructure at hydropower sites,
- Proximity to demand centers in case of water supply reservoirs or lakes,
- Improved energy yield from the cooling effect of water, reduced shading effects, and decreased presence of dust,
- Elimination of the need for large-scale site preparation,
- Additional capacity for hydro plants which increases revenue and simultaneously allows water use for irrigation or other purposes.

The most relevant dynamic modeling and analysis works on a PV + battery + hydro hybrid can be found in Wang and Xu [10], where, to test the stability of the PV hydro system, a dynamic model of a real system was created and simulated with disturbances. The purpose of the collocated PV + battery plant was to supplement the hydrogeneration during the dry season.

The Longyangxia Hydropower plant in Qinghai, China is an example of such a plant with a hybrid hydropower/PV system [11]. On a typical day, the output from the hydro facility is reduced, especially from 11:00 a.m. to 4:00 p.m., while PV generation is high. The saved energy is then requested by the operator for use during early morning and late-night hours. Although the daily generation pattern of the hydropower has changed, the daily reservoir water balance could be maintained at the same level as before to meet the water requirements of other downstream reservoirs.

Gevorgian and O'Neill [12] concluded that, with the fast and precise action of the PV inverters, multiple services such as fast frequency response and power oscillation damping can be provided to the grid, and that the technology required is already available. However, the accurate estimation of available peak power is important for all services, and the extent of these services will depend upon the maximum amount of available PV. When PV penetration is low, there will be other plants in the system that will provide ancillary services; however, when PV penetration increases, it becomes essential that utility-scale PV plants provide these services. The estimation of maximum power available and the DC operating voltage required for a particular power output is of utmost importance.

Hoke, Muljadi, and Maksimovic [13] proposed a second-order polynomial model to estimate the maximum power of a PV module from measured irradiance and temperature, where the coefficients of the model were derived through a linear regression of the PV MPP in different insolation and temperature conditions. Although the model seems accurate when the insolation and temperature is steady, during transient states, the model becomes highly inaccurate. Following the same trend, Hoke et al. [14] stored the P–V curves of the array as a look up table where interpolation is performed with a spline, which requires a considerable amount of memory that introduces error. Pappu, Chowdhury, and Bhatt [15] and Watson and Kimball [16] used pseudo-MPP by controlling the operating voltage to a value greater than the MPP voltage. Here, the operating point is searched by the Perturb and Observe (P&O) algorithm over the power–voltage (P–V) characteristics, implemented through the DC/DC converter. Neely et al. in [17] used the curtailed PV power to realize the primary frequency control using the fixed-droop characteristics.

This paper is organized as follows: Section 2 describes the modeling and control effort for the PV and TPSH systems while also describing the plant controls. Section 3 describes the simulation results of the control algorithms.

2. Modeling and Control

2.1. The Hybrid Plant

A PV plant rated 100 MVA with four PV arrays of 25 MVA each was constructed in Simulink. Each 25 MVA array was assumed to have a lumped DC/DC converter and a lumped DC/AC inverter. A 100 MVA TPSH dynamic model was constructed in Simulink. The TPSH and PV plant were coupled together to the grid using a 110 MVA step-up transformer, as shown in Figure 2.

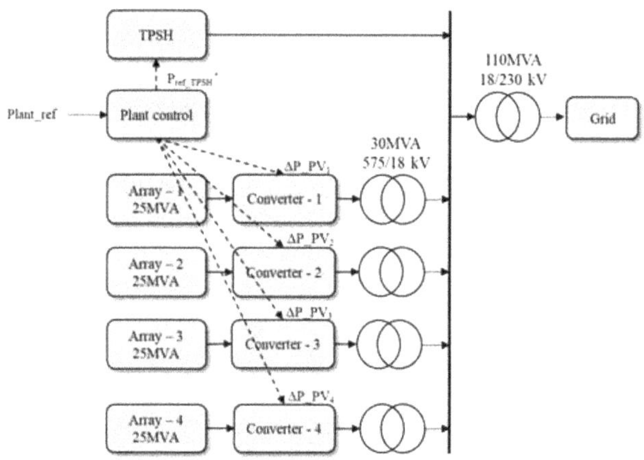

Figure 2. Schematic showing a hybrid PV + TPSH plant.

2.2. Hybrid Pant Control

The plant control, as shown in Figure 3, performs the secondary control which involves set-point distribution among the resources of the hybrid plant.

$$P_{ref\_TPSH} = Plant_{ref} - \sum_{i=1}^{N} P_{PV_i}, \tag{1a}$$

$$P_{ref\_TPSH}^{*} = \begin{cases} \underline{PG}_{TPSH}, & P_{sch} > 0, \quad Pref\_TPSH < \underline{PG}_{TPSH} \\ \overline{PG}_{TPSH}, & P_{sch} > 0, \quad Pref\_TPSH > \overline{PG}_{TPSH} \\ \underline{PP}_{TPSH}, & P_{sch} < 0, \quad Pref\_TPSH < \underline{PP}_{TPSH} \\ \overline{PP}_{TPSH}, & P_{sch} < 0, \quad Pref\_TPSH > \overline{PP}_{TPSH} \end{cases}, \tag{1b}$$

where P_{ref_TPSH}, $P_{ref\_TPSH}^{*}$, $Plant_{ref}$, and P_{sch} stand for the real power reference for TPSH, constrained real power reference for TPSH, reference to the entire plant, and scheduled power for the TPSH; $\underline{PG}_{TPSH}$, $\overline{PG}_{TPSH}$, $\underline{PP}_{TPSH}$, and $\overline{PP}_{TPSH}$ stand for the minimum and maximum possible generation from TPSH, and minimum and maximum possible pump power consumption of TPSH.

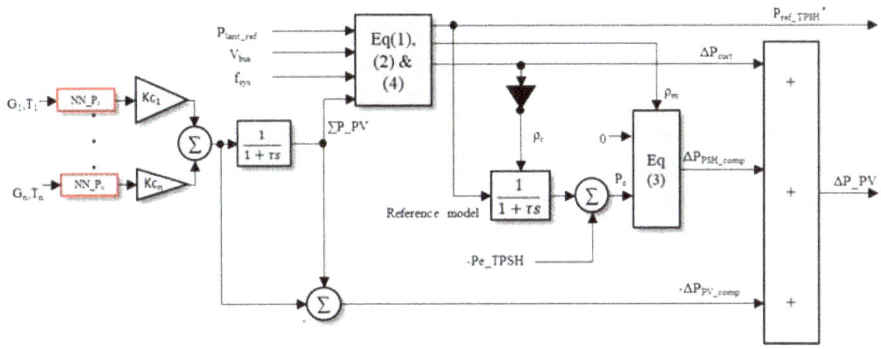

Figure 3. Schematic of the hybrid PV + TPSH plant control.

The plant control estimates the PV generation using NN_P and the array's curtailment gain Kc_n, where NN_P is a neural network emulating the maximum output power of each array (as the output of the NN_P) given the solar irradiance, G, and temperature, T, as inputs. As shown in Figure 3, the plant control utilizes the PV plant reserve and control capability for (a) compensating for sudden changes in insolation through ΔP_{PV_comp}, (b) compensating for hydro plant nonlinear behavior through ΔP_{PSH_comp} and, (c) curtailing the PV plant through ΔP_{curt}. The TPSH is used mainly for (a) firming slow disturbances in PV output, and (b) generation shifting. Both PV and TPSH system participate in voltage control. Some of the design features of the plant control are as follows:

- Each i-th array, producing P_PV_i, is assumed to be curtailed to 10% of its MPP or Kc_n is taken as 0.9 for all N arrays. The 10% curtailment assumption was inspired by [18] which stated that this curtailment is economically compensated for by using the reserve so provided in ancillary services.
- The plant control implements a reference model-based control where a first-order reference model is used to describe the desired behavior of the hydro plant. The PV plant compensates for the difference between desired and actual plant behavior using the ΔP_{PSH_comp} channel through P_c. More details on model reference adaptive control can be found in [5,19].
- The above feature can also lead to the PV plant compensating for the inertia response and primary frequency response of the hydro plant. The logic flags, mode flag ρ_m and reset flag ρ_r, are used to stop compensating for these TPSH plant behaviors in the case of a network disturbance. The flags then reset the model and initialize it to post-disturbance conditions.

$$\rho_m = \begin{cases} 1, & V_{min} < V < V_{max} \text{ and } f_{min} < f < f_{max}, \\ 0, & otherwise \end{cases} \tag{2}$$

$$\Delta P_{PSH\_comp} = \begin{cases} 0, & \rho_m = 0 \\ P_c, & \rho_m = 1 \end{cases}, \tag{3}$$

where, V_{min}, V_{max}, f_{min}, and f_{max} refer to the minimum and maximum limits of terminal voltage V and frequency f, respectively.

- Since NN_P is used by the array control (described in the next section) and by the plant control, any steady-state error in PV output estimation through the ∑P_PV channel is compensated for by the TPSH. This ensures accurate tracking of plant references.
- Curtailing the PV plant is achieved through ΔP_{curt}.

$$\Delta P_{curt} = \begin{cases} del\_PV, & del\_PV > 0 \\ 0, & del\_PV < 0 \end{cases}, \tag{4}$$

where *del_PV* stands for deviation of total generation of the hybrid plant from the plant's reference.

- The PV plant compensates for the difference between desired and actual plant behavior using the ΔP_{PSH_comp} channel through P_c.

To enable the implementation of the described control, Equations (1)–(4) were used in the software defined controller, as shown in Figure 3.

2.3. PV Plant Model and Control

The dynamic model of a solar module can be formulated as a current source with series and shunt resistance to mimic the internal voltage drop and heating due to leakage current. The dynamic model is popularly described as shown in Figure 4. The dynamic model of solar module consists of the photo current or the light-generated current I_{ph}, the dark current or reverse saturation current I_d, and the leakage current I_{sh}.

$$I_{PV} = I_{ph} - I_d - I_{sh}. \tag{5}$$

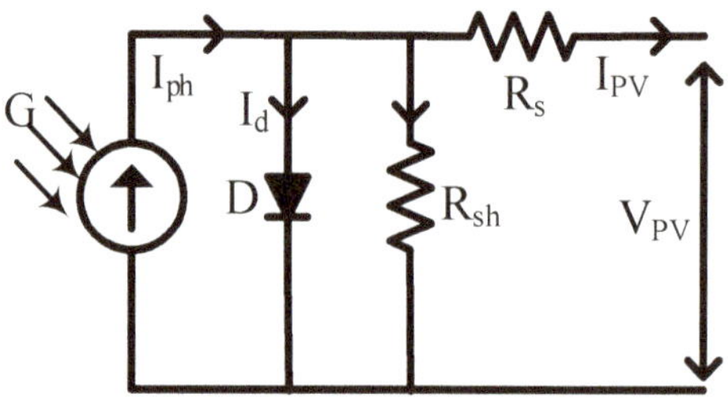

Figure 4. Solar cell model.

The current of a solar module, as a function of voltage, is given as follows [13,20]:

$$I_{ph} = G\frac{I_{sc}[1 + K_i(T - T_n)]}{G_n} - I_o\left(e^{V_D/V_t} - 1\right) - \frac{V_D}{R_{sh}}, \tag{6}$$

where I_{sc} and I_o are the short-circuit and the diode reverse saturation currents, respectively, V_t is the thermal voltage, R_{sh} is the parallel resistances, and V_D is the diode voltage of the cell. K_i is the temperature coefficient of cell's short-circuit current, T is the cell's absolute temperature, T_n is the reference temperature, and G and G_n are the solar radiation at actual and nominal level, respectively. The diode reverse saturation current, I_o, used in Equation (6) is given by

$$I_o = \frac{I_{ph} - V_{oc}/R_{sh}}{e^{\frac{qV_{oc}}{knT}} - 1}. \tag{7a}$$

An alternative and more accurate formulation of Equation (7a) can be given as follows [20]:

$$I_o = \frac{I_{sc}}{e^{\left(\frac{V_{oc}}{V_t}\right)} - 1}\left(\frac{T}{T_n}\right)^3 e^{\left(\frac{E_g}{V_t}\right)\left(\frac{1}{T_n} - \frac{1}{T}\right)}, \tag{7b}$$

where V_{oc} is the cell's open circuit voltage, E_g is the band gap energy of the semiconductor used in the cell, and V_t is the thermal voltage, given by

$$V_t = \frac{\alpha k T}{q}, \tag{8}$$

where q is the electric charge (1.6×10^{-19} C), a is the cell idealizing factor (1.1), and k is the Boltzmann constant (1.38×10^{-23} J/K). The diode voltage V_D can be expressed as

$$V_D = V_t + I_{PV} R_S. \tag{9}$$

To form the array, the generated current and voltage are multiplied by the number of modules in a string or in series and the number of such parallel strings, respectively. The three subsections below describe the control and implementation of array and inverter control. The parameter values of the above equations can be found in Table A4 (Appendix C).

2.4. Array Control through DC/DC Boost Converter

The array controller, as displayed in Figure 5, provides primary frequency support through fixed droop and virtual inertia control. It also tracks the required set-point for the array and implements the commands from plant control. It accomplishes these tasks by controlling the DC/DC boost controller to which the array is connected, as can be seen later in Figure 6.

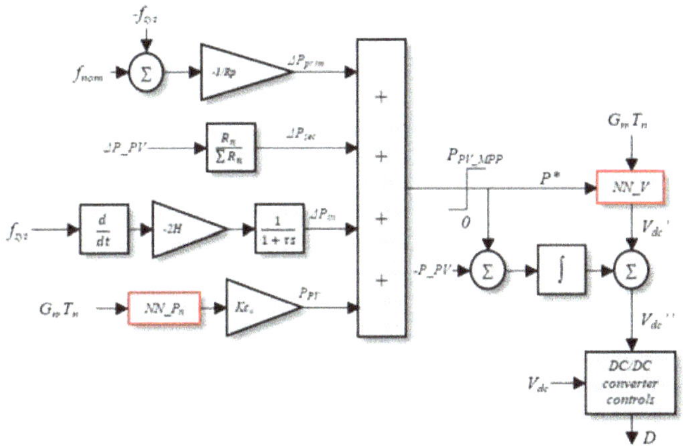

Figure 5. Array control with fixed droop, secondary control, virtual inertia control and curtailment control.

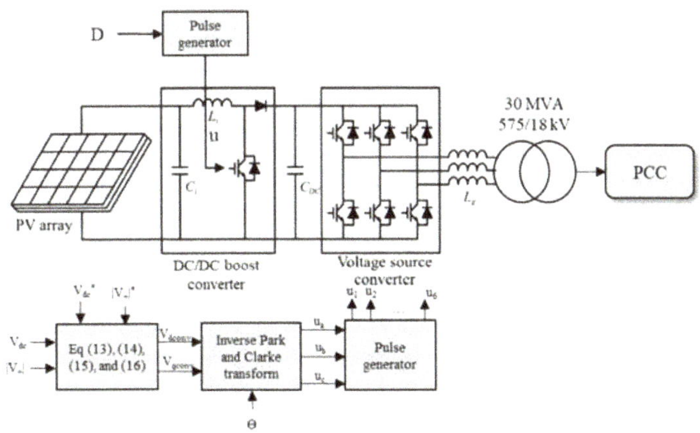

Figure 6. Array control through DC/DC boost converter and inverter control.

The signal ΔP_PV, from plant control, as shown in Figure 3, is shared among the arrays by the ratio of reserve power from each array to the total amount of reserve of the entire plant. Certain arrays of the plant may be completely or partially shaded and, hence, their temperatures will also be different, as will their power output and reserve levels. Thus, the set-point from plant control is divided among the arrays in proportion to their reserves such that the array with largest amount of reserves participates more. This generates the signal ΔP_{sec}, as shown in Figure 5, using

$$\Delta P_{sec} = \frac{\Delta P\_PV\, R_n}{\sum_{n=1}^{N} R_n} = \frac{\Delta P\_PV(1 - Kc_n)NN\_P_n(G_n, T_n)}{\sum_{n=1}^{N}(1 - Kc_n)NN\_P_n(G_n, T_n)}, \quad (10)$$

where R_n is the reserve of array n. The array control also consists of the conventional droop control to produce ΔP_{prim}. The inertia control is formulated as the rate of change of frequency (ROCOF)-based controller, as given below, to produce ΔP_{in}.

$$\Delta P_{in} = -2H_{syn}\frac{df}{dt}, \quad (11)$$

where H_{syn} is the synthetic inertia coefficient, and f is system frequency. Adding inertia, droop, and plant control to the desired steady-state curtailed output P_{PV} produces the real operating point P^*, where P^* is limited by the MPP and 0.

This real operating point P^*, along with the average irradiance over n-th array G_n and array temperature T_n, is used as input to NN_V, to produce Vdc' as the output of NN_V. This DC voltage is further corrected with the help of an integrator to remove NN error and produce Vdc'', which is tracked by the DC/DC boost converter. For this paper, each array is assumed to be curtailed to 10% below its MPP unless purposely curtailed further. The terminal voltage of the PV array is controlled using DC/DC boost converters, as illustrated in Figure 6. The neural networks used for array control are described in the next subsection.

2.5. NN_P and NN_V Training and Implementation for Array Control

First, the MPPs were regressed as a function of irradiance (G) and temperature (T) with NN_P. Three methods of regression were compared for this: (a) curve fit [21], (b) feed-forward neural network (FFNN) [22], and (c) cascade-forward neural network (CFNN) [23]. As can be seen in Table 1, using the curve fit tool resulted in high MSE, while CFNN resulted in the lowest MSE with 15 neurons; hence, the CFNN was adopted for implementation. This produced NN_P (G, T), as described in Equation (10).

Table 1. Mean square error (MSE) for FFNN, CFNN, and curve fit tool for NN_P for MPP regression.

NN_P	FFNN MSE	CFNN MSE	Curve-Fit MSE
3 neurons	9.19×10^{-4}	2.1×10^{-4}	
9 neurons	7.32×10^{-5}	7.76×10^{-5}	0.0185
15 neurons	1.01×10^{-4}	6.5×10^{-5}	

Second, the terminal voltage of the module was regressed for any given power output P^*, irradiance (G) and temperature (T) with NN_V. Figure 7 highlights the P–V characteristics of a single module, where the right-hand side (RHS) of the MPP was regressed. With different number of hidden neurons, the FFNN and CFNN were examined in terms of their regression accuracy, and results of the comparison are shown in Table 2.

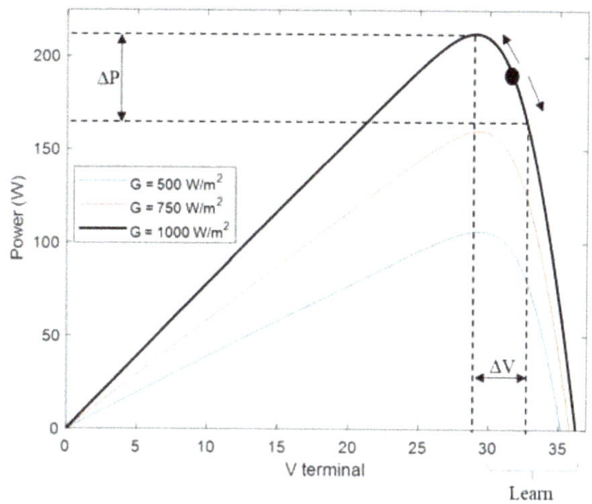

Figure 7. Normalized P–V characteristics of the PV panel showing operating point for T = 25 °C.

Table 2. MSE for FFNN and CFNN for NN_V for V_{dc}' regression.

NN_V	FNN MSE	CFNN MSE
3 neurons	2.56×10^{-2}	1.3×10^{-2}
9 neurons	2.8×10^{-3}	3.2×10^{-3}
15 neurons	8.1×10^{-4}	9.7×10^{-4}

Learning the right-hand side (instead of left-hand side) of the MPP was done for three reasons: (a) to increase regression accuracy of the NN_V (learning both sides of the P–V characteristics gives two voltages for the same power level that reduces regression accuracy), (b) maneuvering the operating voltage in the second half of the MPP requires reduced controller effort and, hence, increased performance, and (c) the DC/DC converter implemented before the inverter is generally a boost converter and, having a higher input voltage, provides headroom for the boost converter. The output of NN_V (*G*, *T*, *P**) is the voltage for the desired power output or the voltage setpoint, *Vdc*′.

Here, two separate networks were used to generate *P** and V_{dc}' to maintain a high level of accuracy. Both networks were trained with the Levenberg–Marquardt algorithm [24,25].

To train NN-V (*G*, *T*, *P**), 112,477 samples (*G*, *T*, and desired power *P** as inputs and Vdc′ as the output) were extracted from the P–V characteristics. The model was simulated under different operating conditions of *G*, *T*, and *P** to collect V_{dc}'. These samples were randomly divided such that 70% were used for training, 20% were used for validation, and 10% were used for testing. To train NN_P (G, T), 1911 samples of MPP were used and were divided as above. The samples consisted of *G* and *T* as inputs and maximum power as the output, and they were collected by simulating the model under different conditions of *G* and *T*. The error frequency distribution results of the NNs from testing are displayed in Appendix D.

2.6. Inverter Control

The array control can accommodate the primary and secondary control but will perturb the DC voltage. To maintain the DC voltage and the AC bus voltage, the inverter control described below is used. The PV system is interfaced to the grid with a DC/DC boost converter and an inverter with an inductive filter. The inductive filter and its inherent DC resistance can modeled as

$$\begin{bmatrix} \dot{i_d} \\ \dot{i_q} \end{bmatrix} = \begin{bmatrix} -R_f/L_f & \omega_s \\ -\omega_s & -R_f/L_f \end{bmatrix} \begin{bmatrix} i_d \\ i_q \end{bmatrix} + \begin{bmatrix} 1/L_f & 0 \\ 0 & 1/L_f \end{bmatrix} \begin{bmatrix} v_{dg} - V_{dconv} \\ v_{qg} - V_{qconv} \end{bmatrix}. \quad (12)$$

R_f and L_f are filter resistance and inductance, respectively. The parameter values can be found in Table A5. The DC voltage reference, ${V_{dc}}^*$, and positive sequence bus voltage magnitude reference, $|V_+|^*$, can be tracked if we choose the control V_{dconv}, and V_{qconv}, as follows:

$$V_{dconv} = V_{dcomp} + PI_1(PI_2({V_{dc}}^* - V_{dc}) - i_d), \quad (13)$$

$$V_{qconv} = V_{qcomp} + PI_1(PI_2(|V_+|_* - |V_+|) - i_q), \quad (14)$$

where PI_1 is a proportional and integral controller that forms the inner loop and tracks the dq-axis current, whose reference is generated by the PI_2, where PI_2 is a proportional and integral controller that forms the external loop and tracks the DC voltage or the positive sequence bus voltage magnitude.

$$V_{dcomp} = v_{dg} + i_q \omega_s L_f \quad (15)$$

$$V_{qcomp} = v_{qg} - i_d \omega_s L_f \quad (16)$$

By choosing V_{dcomp} and V_{qcomp}, we can cancel out the cross-coupling of states, i_q and i_d, in Equation (12). The *PI* controllers can be tuned using transfer function approximations of the compensated filters and pole placement techniques. Here, v and V indicate measured and controlled voltage. Subscripts *d*, *q*, *comp*, *conv*, and *g* stand for *d*-axis, *q*-axis, compensation, converter-side values, and grid-side values; ω_s is the synchronous frequency in rads/s.

2.7. TPSH Plant Model and Control

A 100 MW TPSH dynamic model was constructed in Simulink with parameters listed in Table A3. The TPSH turbine and governor model was constructed as described in previous works and studies [7,26] with a few changes to accommodate the logic flags for automatic mode change. This model is available in Figure A1. The automatic mode change control allowed the TPSH to change its modes simply on the basis of the desired power reference. To further detail the automatic mode change control, the pseudocode for the mode change is provided in Appendix B. According to this pseudocode, the controller generates logic flags ρ_p, and ρ_t , with which it changes the mode of operation of the TPSH. The logic flags indicate on or off commands to the pump and turbine, respectively. The synchronous machine used for the TPSH is a sixth-order machine model [27], with a standard IEEE Type1 excitation system [28], without any stabilizer. All these components are readily available in Simulink libraries. The parameters used for all these components are provided in Tables A1–A3. The model was implemented in [7,26], which demonstrated that it can describe the reference tracking performance, the fast frequency response performance of the TPSH in both turbine and pump mode, and its mode change capability.

3. Results

3.1. Implementation

The important model modifications and techniques used while implementing the model shown in Figure 3 were as follows:

- Although the TPSH mode change capability can be used in firming, it was avoided. This was done mainly to avoid repeated mode change and oscillation between modes, which can lead to early failure of the ternary set.
- The excitation system of the TPSH and inverters maintain the voltage set or determined by the load flow performed before the initialization.

- The TPSH model implemented head by dividing the discharge by the gate command. The gate command is limited to a minimum value of $G_{min} > 0$. Thus, the TPSH model would never have a 0 output for mechanical and electrical power.
- The reverse saturation current depends on the photo current, which created an algebraic loop. To break this algebraic loop, a unit delay was used.
- Through MATLAB code, the neural network toolbox of MATLAB allowed the training of the NNs and deployment of the NNs into Simulink, with a particular sample time suited for the simulation.
- To avoid a time-varying filter model, ω_s in Equation (12) was fixed at ω_o with 60 Hz as the nominal frequency.
- For the simulation of the 9-bus system in pump mode with HSC and pure pump mode, real and reactive power loads at buses 5, 7, and 9 were reduced to 50% for HSC mode and 10% for pure pump mode.

3.2. PV Plant Set Point Tracking

To show the accuracy and efficiency of the PV plant and array control, the PV plant with four 25 MW arrays was integrated to an infinite bus and tested for its set-point tracking capability. The main aim of performing set-point tracking is to validate the control scheme described in Figure 5. The solar irradiation levels for the four PV arrays were set to 850, 950, 900, and 400 W/m^2 with temperature set to 25 °C. Step increments of power reference were made at intervals of 5 s to replicate automatic generation control (AGC) signals. If all arrays were curtailed by 10% with insolation levels being 850, 950, 900, and 400 W/m^2, array 4 would have the least amount of reserves. When the AGC step change was equally divided, array 4 was not be able to supply its share accurately as array 4 would saturate at its own MPP faster than the others and, hence, would cause inaccurate tracking of the set-point, as we can see at t = 7, 11, and 24 s in Figure 8a. However, when the set-point change was divided in proportion to the reserve of each array, accuracy of set-point tracking increased, as shown in Figure 8b. In other words, the controls were able to coordinate unshaded and shaded arrays to meet a particular set-point.

It was also noticed that the set-point tracking with neural networks was highly accurate. From Figure 8, the output was tracked accurately when the desired output was below the MPP. However, if the desired output was greater than the MPP of the plant, the output saturated at the MPP, as observed from t = 11 to 24 s. Figure 8c reflects the accurate and smooth performance of the inverter controls.

3.3. Hybrid Plant Set Point Tracking

Although the PV plant can perform set-point tracking, it can only do so within certain limits of available PV power. Thus, the reserve provided by the PV system was used for primary frequency response and TPSH response enhancement only. When the set-point for the hybrid plant changes, the PV plant momentarily responds to correct the error between the hydro plant and the reference model as dictated by the plant control system in Figure 3.

In Figure 9b,d,f, at t = 10 s, when the reference increased, the PV system was first to respond and then attempted to absorb the reverse response of the hydro turbine governor system. This capability decreases the response time of the plant and prevents the hydraulic nonlinearities from interfering with the grid, as can be seen in Figure 9a,c,e. However, the PV plant did not compensate for the tracking error completely due to the delays in the inverter control. Moreover, due to its nonlinear nature, the TPSH model exhibited greater oscillations on set-point reduction, which resulted in an increase in transient set-point tracking error. Nonetheless, this capability also provided the plant with pump mode flexibility, as can be seen in Figure 9e.

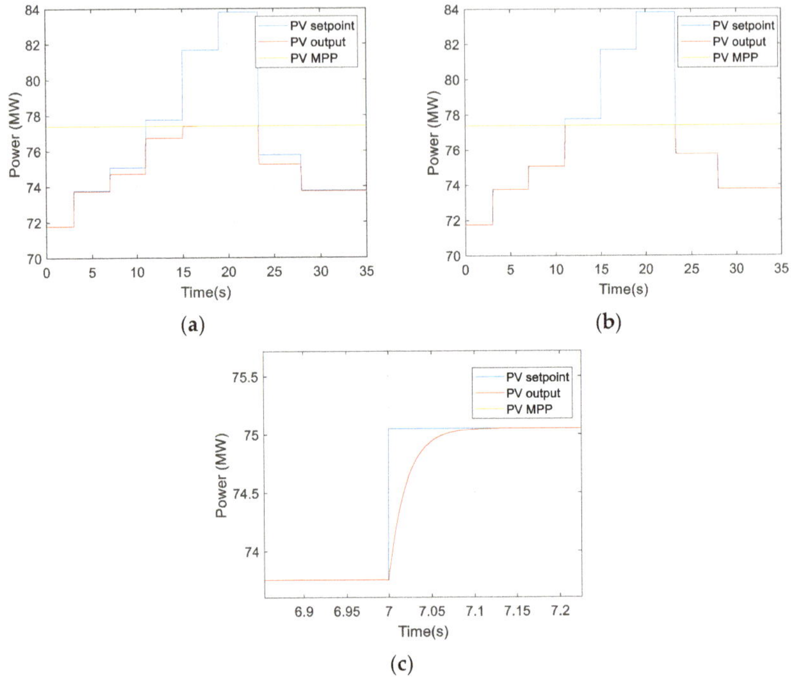

Figure 8. Plant control capability when the (**a**) reference was equally divided, and (**b**) reference was divided proportional to the reserve; (**c**) inverter control performance.

3.4. PV Firming Results

TPSH can perform firming only in generation mode and pump mode with HSC active. While mechanical governors cannot be used to compensate for rapid changes in PV, they can be used for slow changes in PV. A low-pass filter is used to filter out the fast transients in the PV output as shown in Figure 3, while the self-compensation feature through ΔP_{PV_comp} of the PV inverter controls is used for compensating fast changes in PV and for the damping transients due to mode change from pump to generation mode of the TPSH. Figures 10–12 show a case study for one day. To ease the simulation burden, certain sections of the day were picked and simulated separately.

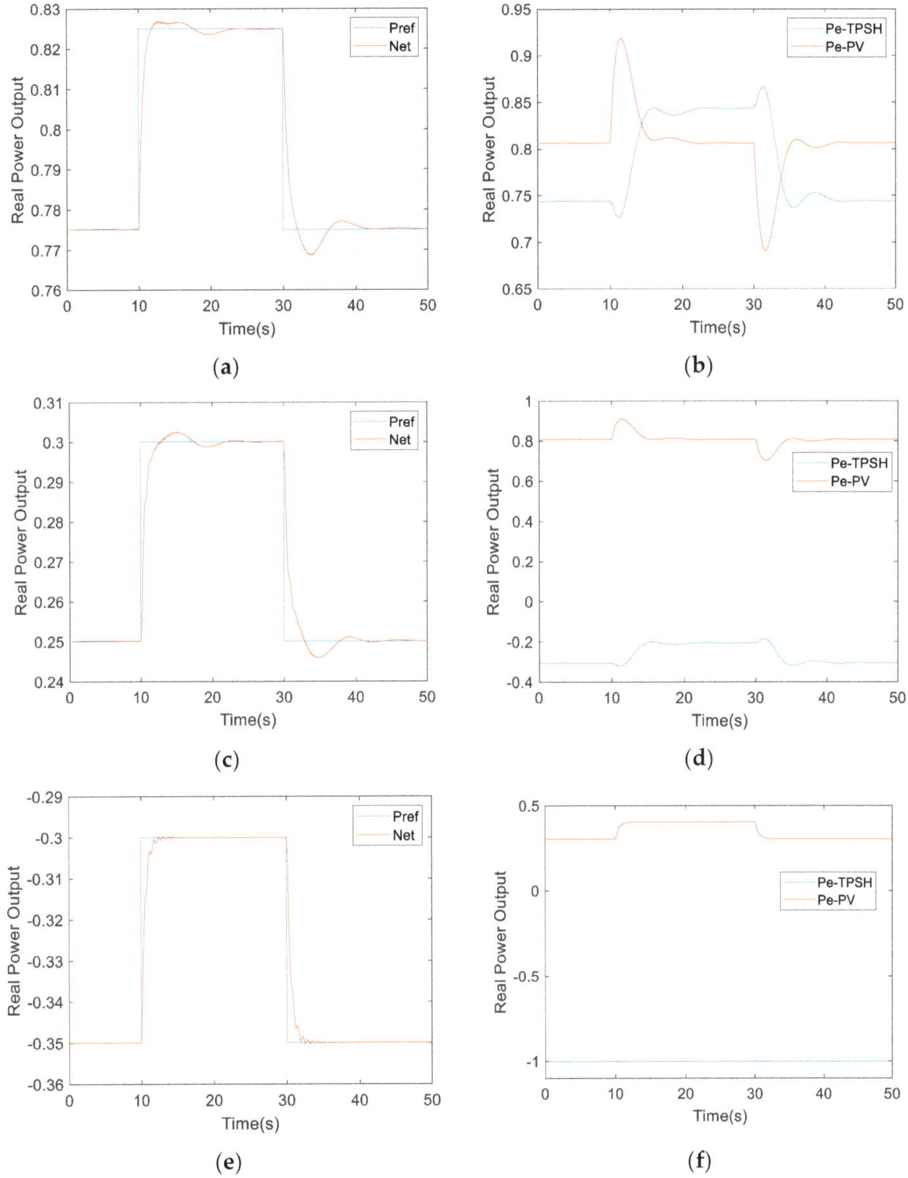

Figure 9. Set-point tracking capability and individual power output of the PV and TPSH plant in (**a**,**b**) generation mode, (**c**,**d**) pump mode with hydraulic short-circuit, (**e**,**f**) pure pump mode.

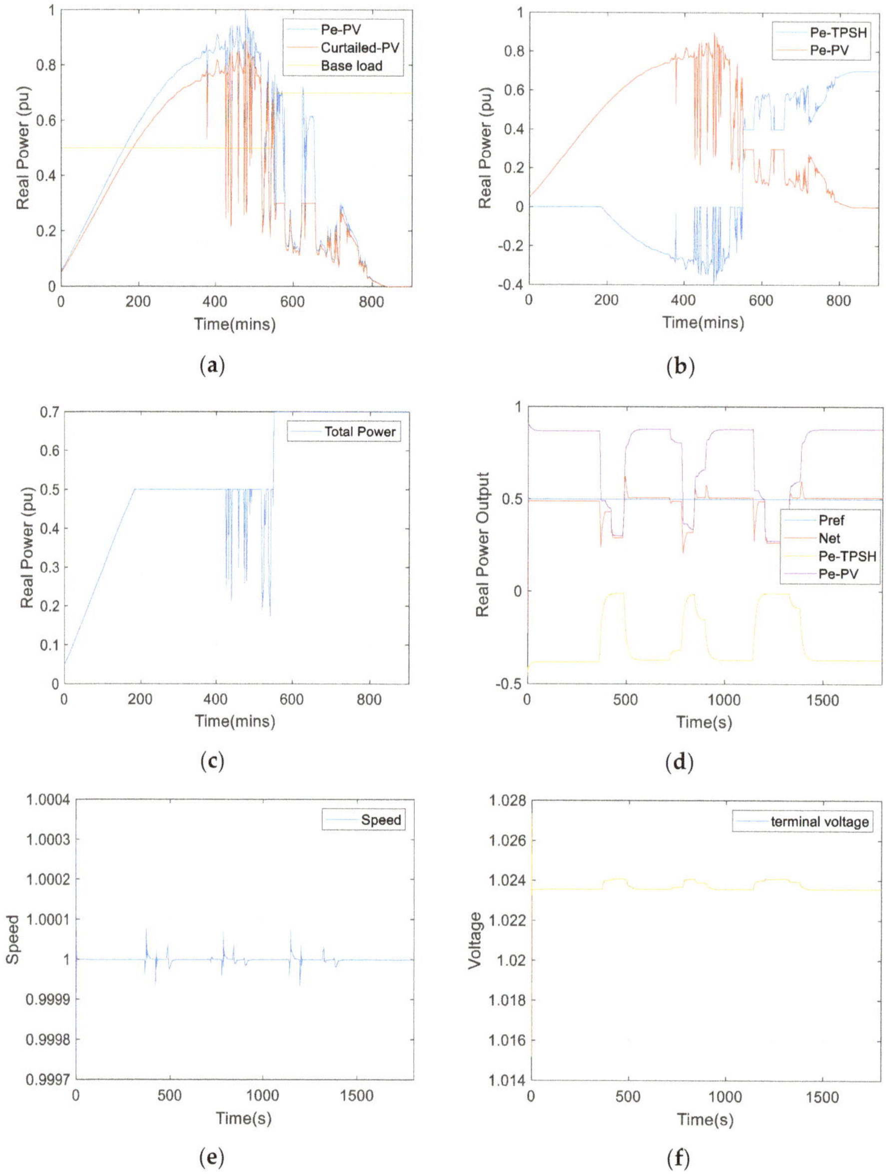

Figure 10. PV firming from t = 420 min to t = 450 min showing (**a**) irradiance profile, (**b**) required TPSH power to meet baseload, (**c**) firmed PV power (**d**) power output from PV and TPSH during dynamic simulation to meet set-point (**e**) speed and (**f**) terminal voltage during simulation.

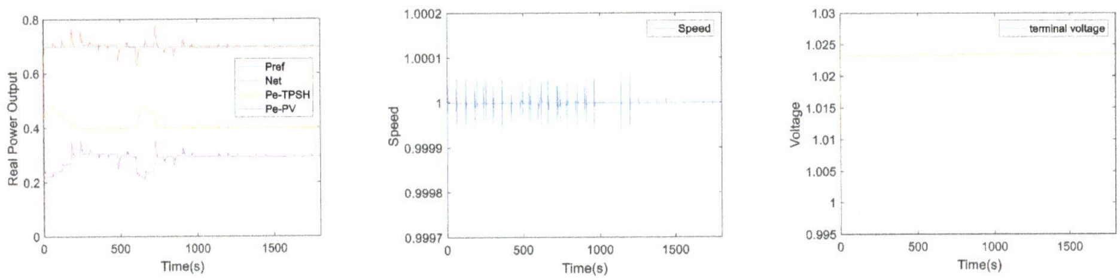

(a) (b) (c)

Figure 11. PV firming from t = 610 min to t = 640 min showing (**a**) power output from PV and TPSH during dynamic simulation to meet set-point (**b**) speed and (**c**) terminal voltage during simulation.

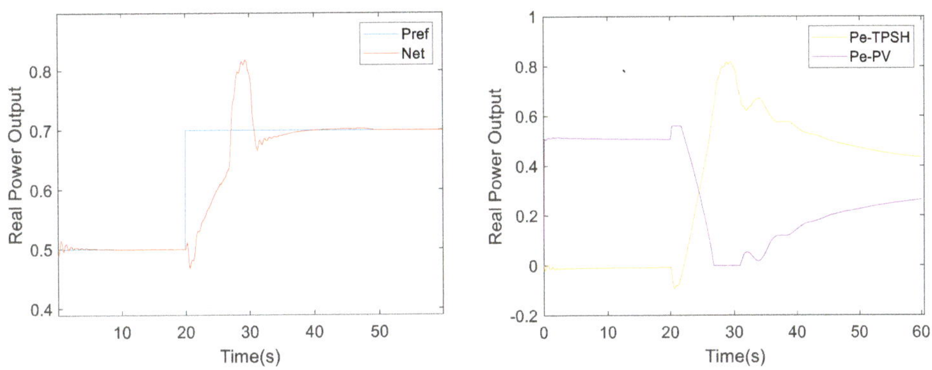

(a) (b)

Figure 12. Transition firming showing (**a**) net response of the hybrid plant and (**b**) power output from PV and TPSH during dynamic simulation to meet set-point.

The steady-state simulation of the day is shown in Figure 10a–c. In Figure 10a, we see the PV profile with and without curtailment and the base load. The aim of the controls was to modify Figure 10a into Figure 10c using actions in Figure 10b. The base load in the morning hours was ignored, and the plant was allowed to ramp up to the base load, which would support load increase. Figure 10d–f display the dynamic simulation results for a period of 30 min (from t = 420 min to t = 450 min) of the profile in Figure 10a. During this period, the TPSH operated in pump mode with HSC and was seen to adjust its output to minimize the error in set-point tracking. By operating in pump mode with HSC, it prevented curtailment of the PV system. However, since the TPSH was prevented from transitioning into the generation mode and due to reserve limitations, it could not reduce the plant's tracking error to zero.

Figure 11 shows the firming effect of the controls using the generation mode of the TPSH. Due to the availability of upward reserve, the set-point was tracked with small transient errors. As shown in Figure 10a, the TPSH changed its mode from HSC to generation, as scheduled to meet the increase in base load from 50% to 70%. The dynamic simulation of this phase is shown in Figure 12. The TPSH had a minimum generation of 40%, and the available PV was 50%. The controls curtailed the PV system to compensate for the transition transients of the TPSH and meet the base load of 70%. The PV system saturated and was unable to completely compensate for the transition transients.

3.5. Fast Frequency Response of Hybrid Plant in IEEE 9-Bus System

The above tests verified the secondary control capability of the hybrid plant and its controls. The subsequent tests were performed to verify the primary controls. The hybrid plant was integrated to the IEEE 9-bus system [29], as indicated in Figure 13, and a load was applied suddenly at bus 5 at t = 60 s. As a result, the frequency of the system declined. The frequency profile for this test is shown in Figure 14a,c,e, where the TPSH is in generation mode, pump mode with HSC active, and pure pump mode. The real power contributions from the hybrid and TPSH plant are compared in Figure 14b,d,f. In all cases, the hybrid plant outperformed the TPSH plant in that it improved the frequency nadir and damped frequency oscillations. This was mainly attributed to the joint response of the PV and TPSH plant. The PV plant utilized its reserves for inertia control and droop control, to support frequency.

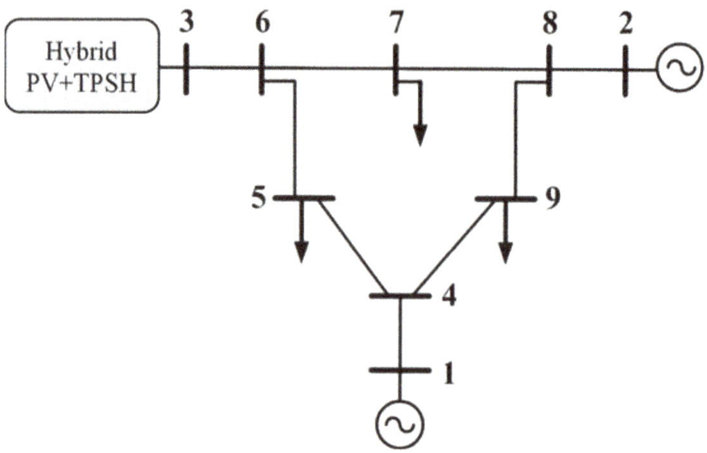

Figure 13. IEEE 9-bus system with proposed PV plant and controls at bus 3.

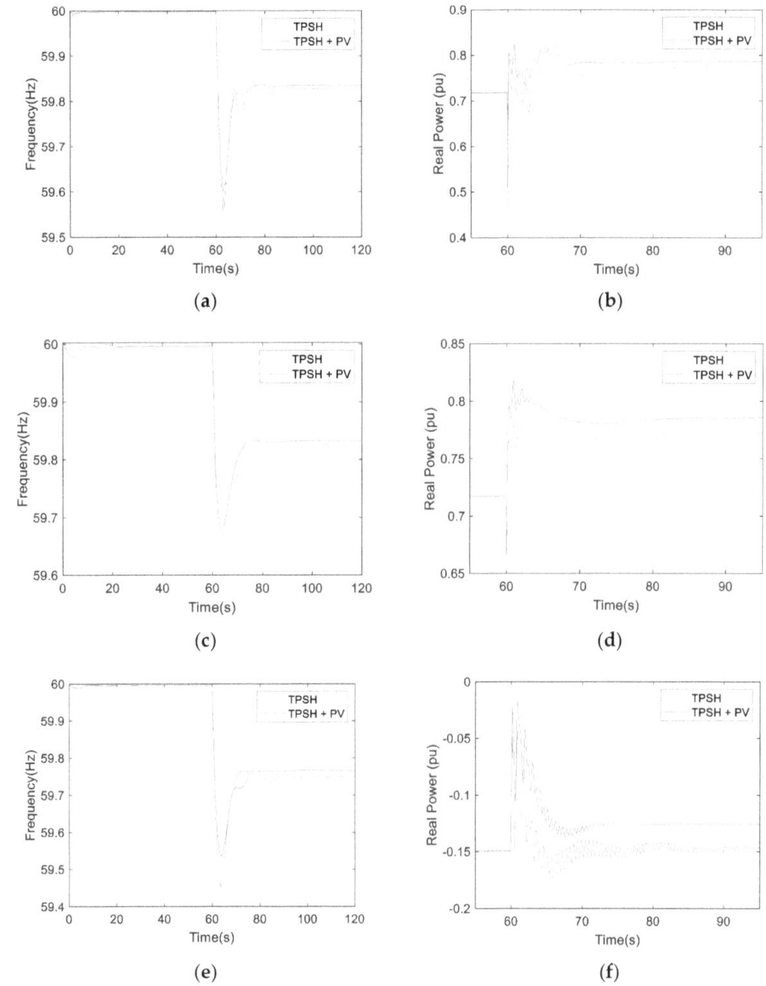

Figure 14. FFR results of frequency and real power output: (**a**,**b**) generation mode, (**c**,**d**) pump mode with HSC, and (**e**,**f**) pure pump mode.

4. Conclusions

To cope with high flexibility and resilience requirements of future low inertia power systems, this paper proposes a novel PV + TPSH hybrid plant and designs its controls using a neural network-based structure. The TPSH was chosen because of its pump mode flexibility. The aim was to exploit the flexibility of the TPSH and leverage the controller-induced flexibility of the PV system to make the hybrid plant be-have as a conventional plant. The advantages of the proposed hybrid configuration include (a) the addition of generation capacity, (d) the formation of a self-sufficient source, and (d) reduced transmission loss in pump mode. If a floating PV system is used, the additional advantages of that can be realized as well.

The designed controls can be decomposed into three parts: (a) hybrid plant control, (b) PV plant control, and (c) TPSH plant control. The hybrid plant control distributes the set-point among the TPSH and PV systems, while the array controls make sure that the hybrid plant control commands are followed accurately, and that the primary control is provided. The TPSH control is a nominal hydro-governor enhanced with mode change capability. To control the PV arrays, a neural network-based reference governor framework was used, with a mixture of CFNN and FFNN. Here, two separate neural networks are used to accurately regress the required DC link voltage and the maximum power as a function of irradiance and temperature.

Through detailed simulation case studies, it was shown that, using the designed controls within operational constraints, the hybrid plant behaved like a conventional plant. In addition to configuration advantages mentioned above, it is concluded that the controls can (a) enable the PV plant to track set-points, (b) coordinate the response and coordination of shaded and unshaded PV arrays, (c) enhance the response of the TPSH plant for set-point tracking in generation and pump mode with HSC, (d) enable pure pump mode set-point tracking (f) can firm the PV plant with array control, systematic curtailment and TPSH response, and (g) enhance fast frequency response through combined response of PV and TPSH.

Author Contributions: Conceptualization, S.N., and K.Y.L.; methodology, S.N., and K.Y.L.; validation, S.N., and K.Y.L.; investigation, S.N.; writing—original draft preparation, S.N.; writing—review and editing, K.Y.L.; supervision, K.Y.L.; All authors have read and agreed to the published version of the manuscript.

Funding: This research received no external funding.

Institutional Review Board Statement: Not applicable.

Informed Consent Statement: Not applicable.

Conflicts of Interest: The authors declare no conflict of interest.

Appendix A

The parameter values used for TPSH governor turbine, excitation, and synchronous machine can be found in the tables below.

Table A1. Synchronous machine parameters [28].

Pn **(MVA)** ***Vn*** **(Vrms)** ***fn*** **(Hz)**	**[*Xd Xd′ Xd″ Xq Xq″ Xl*] (pu):**	**[*Td′ Td″ Tqo″*] (s):**	***Rs*** **(pu):**	***H*** **(s)** ***F*** **(pu)** ***p*** **(-)**
[100 18,000 60]	[1.305, 0.296, 0.252, 0.474, 0.243, 0.18]	[1.01, 0.053, 0.1]	2.8544×10^{-3}	[3.7 0 32]

Table A2. Exciter parameters [29].

Tr	*Ka, Ta* (s)	*Ke, Te* (s)	*Tb* (s), *Tc* (s)	*Kf, Tf*	*Efmin, Efmax, Kp*
0.02	200, 0.001	1, 0	0, 0	0.001, 1 s	0, 7, 0

Table A3. Governor turbine system parameters referring to Figure A1.

Kg and *Kp*	τg and τp	*Dturb*	*Twt*	*Twp*	*At*	*fp*
1	0.5	0.5	1.605	1.605	0.98	0.0003042
Kp	*Ki*	*Kd*	*Tf*	*Ht* and *Hp*	*Gmax* and *Gmin*	*VGmax* and *Vgmin*
0.5818	0.105	0.015	0.001	1	1 & 0.01	±0.1/s
Qnl	*Rp*	τf and τpe				
0.06	0.05	0.01				

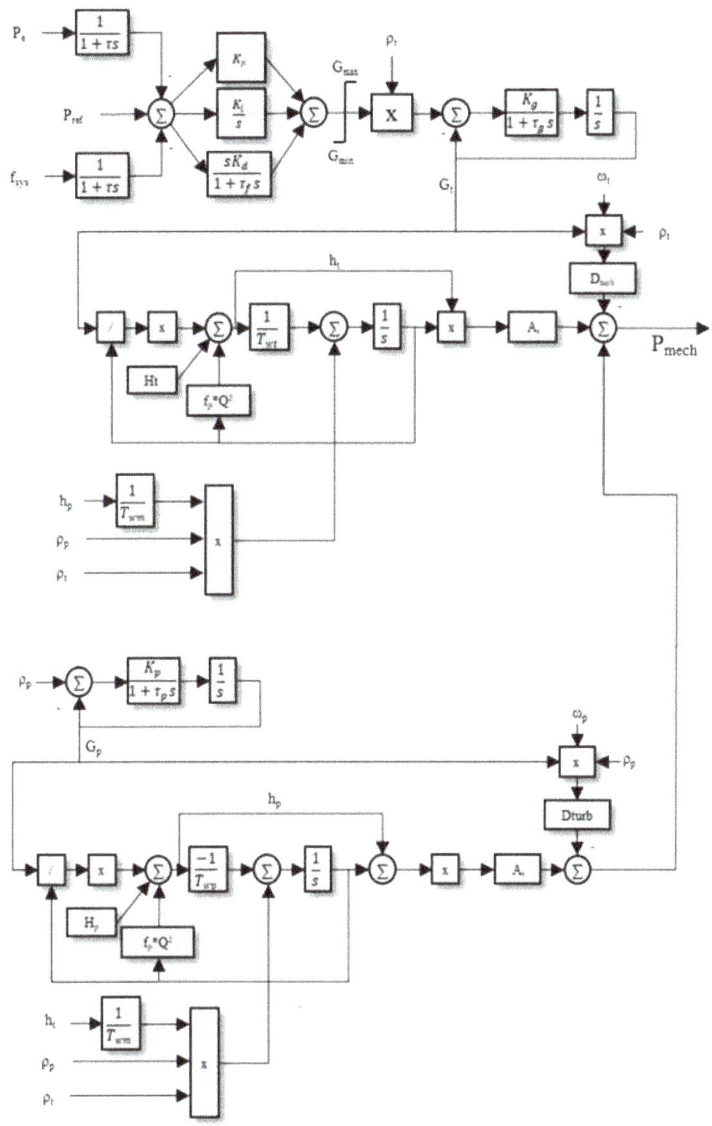

Figure A1. Governor and turbine model for the TPSH.

Appendix B

The pseudo code for the automatic mode change algorithm is described below. Here, the modes 1, 2, and 3 correspond to generation, pump, and pump with hydraulic short-circuit modes; MC indicates the required mode change where, for example, 23 indicates pump to pump with hydraulic short-circuit; ρ_p and ρ_t are binary variables that indicate on/off status for the pump and turbine gates as in Figure A1.

Algorithm A1. Automatic mode change of TPSH

```
if MC is 23
    set ρp as 1 and ρt as 1;
    update Pref as 1+Pref;
    set Present mode as 3;
if MC is 12
    update Pref as 0;
    if Pe is less than εe
        set ρp as 1 and ρt as 0;
        set Present mode as 2;
    else
        set ρp as 0 and ρt as 0;
        set Present mode as 1;
if MC is 13
    set ρp as 1;
    if Pe is less than εe
        update Pref as 1 + Pref;
        set ρt as 1;
        set Present mode as 3;
     else
        set Present mode as 1;
if MC is 21
     set ρp as 0;
     if Pe is greater than -εe
        set ρt as 1;
        set Present mode is 1;
     else
        update Pref as 0;
        set ρt as 0;
        set Present mode as 2;
if MC is 32
     update Pref as 0;
     set ρt as 0;
     set ρp as 1;
     set Present mode as 2;
        if MC is 31
     set ρp as 0;
     update Pref as Pref;
     set ρt as 1;
    set Present mode as 1;
     if Pref is greater than 0
           set ρt as 1;
           set ρp as 0;
           set Present mode as 1;
     elseif Pref is lesser than -.6
           update Pref as 0;
           set ρt as 0;
           set ρp as 1;
           set Present mode as 2;
     elseif Pref is lesser than 0 and Pref is greater than -.6
           update Pref as 1 + Pref;
           set ρt as 1;
           set ρp as 1;
           set Present mode as 3;
```

Appendix C

The parameters for the solar PV modules and inverters are presented below.

Table A4. Solar module parameters [30].

Rsh	*Rs*	*Isc*	*Voc*	*Ncell*
313.3991	0.39383	7.84	36.3	60

Table A5. Inverter and filter parameters.

Lf	*Rf*	DC Link Voltage	*Cdc*	Nominal Power	Phase-to-Phase Vrms
0.15 pu	0.0015 pu	1200	600 uF	25 MVA	575

Appendix D

Error frequency distributions for NN_P are given below.

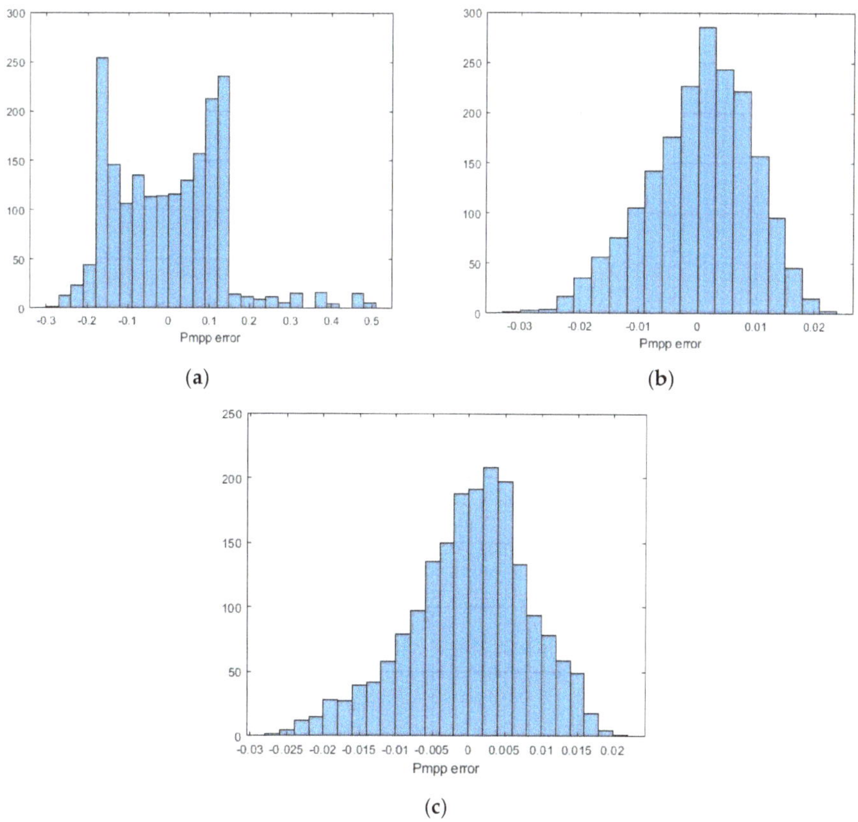

Figure A2. Error frequency distributions with (**a**) curve fit, (**b**) feed forward NN, and (**c**) cascade-forward NN.

Error frequency distributions for NN_V are given below.

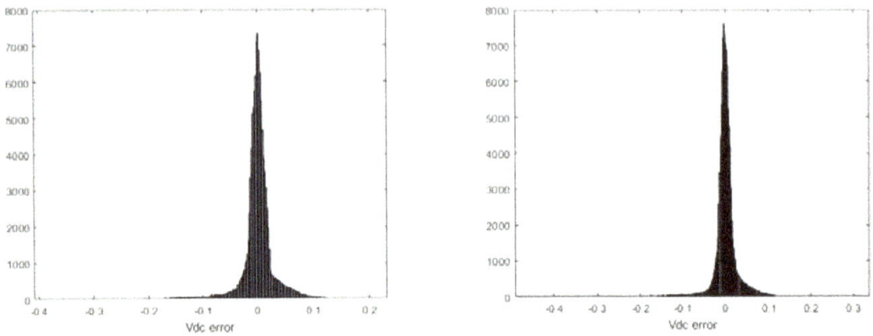

(**a**) (**b**)

Figure A3. Error frequency distributions with (**a**) feed forward NN, and (**b**) cascade-forward NN.

References

1. Center for Climate and Energy Solutions. Hurricanes and Climate Change | Center for Climate and Energy Solutions. 2019. Available online: https://www.c2es.org/content/hurricanes-and-climate-change/ (accessed on 10 July 2021).
2. Energy.gov. 2021. Available online: https://www.energy.gov/sites/default/files/2019/07/f65/Storage%20Cost%20and%20Performance%20Characterization%20Report_Final.pdf (accessed on 10 July 2021).
3. Pdf.directindustry.com. Hybrid Solutions-GE Power-PDF Catalogs | Technical Documentation | Brochure. 2021. Available online: https://pdf.directindustry.com/pdf/ge-power/hybrid-solutions/205105-810039.html (accessed on 22 June 2021).
4. Muljadi, E.; Singh, M.; Gevorgian, V.; Mohanpurkar, M.; Hovsapian, R.; Koritarov, V. Dynamic modeling of adjustable-speed pumped storage hydropower plant. In Proceedings of the 2015 IEEE Power & Energy Society General Meeting, Denver, CO, USA, 26–30 July 2015; pp. 1–5. [CrossRef]
5. Nag, S.; Lee, K.Y.; Suchitra, D.; Mediratta, V. Load Frequency Control through DFIM-based Pumped Storage Hydro. In Proceedings of the 2019 IEEE Power & Energy Society General Meeting (PESGM), Atlanta, GA, USA, 4–8 August 2019; pp. 1–5. [CrossRef]
6. Kratirov, V.; Guzwoski, L. *Modelling Ternary Pumped Storage Units*; Argonne National Laboratory: Oak Ridge, TN, USA, 2013.
7. Nag, S.; Lee, K.Y.; Suchitra, D. A Comparison of the Dynamic Performance of Conventional and Ternary Pumped Storage Hydro. *Energies* **2019**, *12*, 3513. [CrossRef]
8. Dong, Z.; Tan, J.; Muljadi, E.; Nelms, R.M.; St-Hilaire, A.; Pevarnik, M.; Jacobson, M.D. Developing of Quaternary Pumped Storage Hydropower for Dynamic Studies. *IEEE Trans. Sustain. Energy* **2020**, *11*, 2870–2878. [CrossRef]
9. World Bank. Where Sun Meets Water: Floating Solar Market Report. 2021. Available online: https://www.worldbank.org/en/topic/energy/publication/where-sun-meets-water (accessed on 22 June 2021).
10. Wang, Y.; Xu, H. Research and practice of designing hydro/photovoltaic hybrid power system in microgrid. In Proceedings of the 2013 IEEE 39th Photovoltaic Specialists Conference (PVSC), Tampa, FL, USA, 16–21 June 2013; pp. 1509–1514. [CrossRef]
11. Hydropower.org. Case Study: Solar PV–Hydro Hybrid System at Longyangxia, China. 2021. Available online: https://www.hydropower.org/blog/case-study-solar-pv-e2-80-93hydro-hybrid-system-at-longyangxia-china (accessed on 23 June 2021).
12. Gevorgian, V.; O'Neill, B. *Advanced Grid-Friendly Controls Demonstration Project for Utility-Scale PV Power Plants*; Tech. Rep. NREL/TP-5D00-65368; NREL: Golden, CO, USA, 2016.
13. Hoke, A.; Muljadi, E.; Maksimovic, D. Real-time photovoltaic plant maximum power point estimation for use in grid frequency stabilization. In Proceedings of the IEEE 16th Workshop on Control and Modeling for Power Electronics (COMPEL), Vancouver, BC, Canada, 12–15 July 2015; pp. 1–7. [CrossRef]
14. Hoke, A.F.; Shirazi, M.; Chakraborty, S.; Muljadi, E.; Maksimovic, D. Rapid Active Power Control of Photovoltaic Systems for Grid Frequency Support. *IEEE J. Emerg. Sel. Top. Power Electron.* **2017**, *5*, 1154–1163. [CrossRef]
15. Pappu, V.A.K.; Chowdhury, B.; Bhatt, R. Implementing frequency regulation capability in a solar photovoltaic power plant. In Proceedings of the North American Power Symposium, Arlington, TX, USA, 26–28 September 2010; pp. 1–6. [CrossRef]
16. Watson, L.D.; Kimball, J.W. Frequency regulation of a microgrid using solar power. In Proceedings of the Twenty-Sixth Annual IEEE Applied Power Electronics Conference and Exposition (APEC), Fort Worth, TX, USA, 6–11 March 2011; pp. 321–326. [CrossRef]
17. Neely, J.; Johnson, J.; Delhotal, J.; Gonzalez, S.; Lave, M. Evaluation of PV frequency-watt function for fast frequency reserves. In Proceedings of the IEEE Applied Power Electronics Conference and Exposition (APEC), Long Beach, CA, USA, 20–24 March 2016; pp. 1926–1933. [CrossRef]
18. Nrel.gov. 2021. Available online: https://www.nrel.gov/docs/fy16osti/65368.pdf (accessed on 10 July 2021).
19. Astrom, K.; Wittenmark, B. *Adaptive Control Systems*; Addison-Wesley: Reading, MA, USA, 1995.

20. Villalva, M.; Gazoli, J.; Filho, E. Comprehensive Approach to Modeling and Simulation of Photovoltaic Arrays. *IEEE Trans. Power Electron.* **2009**, *24*, 1198–1208. [CrossRef]
21. Mathworks.com. Fit Curve or Surface to Data—MATLAB FIT. 2021. Available online: https://www.mathworks.com/help/curvefit/fit.html (accessed on 8 June 2021).
22. Mathworks.com. Generate Feedforward Neural Network—MATLAB Feedforwardnet. 2021. Available online: https://www.mathworks.com/help/deeplearning/ref/feedforwardnet.html (accessed on 8 June 2021).
23. Mathworks.com. Cascade-Forward Neural Network—MATLAB Cascadeforwardnet. 2021. Available online: https://www.mathworks.com/help/deeplearning/ref/cascadeforwardnet.html (accessed on 8 June 2021).
24. Levenberg, K. A method for the solution of certain problems in least squares. *Q. Appl. Math.* **1994**, *5*, 164–168.
25. Marquardt, D. An algorithm for least-squares estimation of nonlinear parameters. *J. Soc. Ind. Appl. Math.* **1963**, *11*, 431–441. [CrossRef]
26. Nag, S.; Lee, K.Y. Power System Resiliency Enhancement with Ternary Pumped–Storage Hydropower. *IFAC-PapersOnLine* **2020**, *53*, 12714–12718. [CrossRef]
27. Mathworks.com. Model Dynamics of Three-Phase Round-Rotor or Salient-Pole Synchronous Machineusing Fundamental Parameters in pu Units-Simulink. 2021. Available online: https://www.mathworks.com/help/physmod/sps/powersys/ref/synchronousmachinepufundamental.html (accessed on 8 June 2021).
28. Mathworks.com. Provide Excitation System for Synchronous Machine and Regulate Its Terminal Voltage in Generating Mode-Simulink. 2021. Available online: https://www.mathworks.com/help/physmod/sps/powersys/ref/excitationsystem.html (accessed on 8 June 2021).
29. Delavari, A.; Kamwa, I.; Brunelle, P. Simscape Power Systems Benchmarks for Education and Research in Power Grid Dynamics and Control. In Proceedings of the IEEE Canadian Conference on Electrical & Computer Engineering (CCECE), Quebec, QC, Canada, 13–16 May 2018; pp. 1–5. [CrossRef]
30. Mathworks.com. Implement PV Array Modules-Simulink. 2021. Available online: https://www.mathworks.com/help/physmod/sps/powersys/ref/pvarray.html (accessed on 23 June 2021).

energies

MDPI

Article

Investigations of Rake and Rib Structures in Sand Traps to Prevent Sediment Transport in Hydropower Plants

Mads Mehus Ivarson [1], Chirag Trivedi [1,*] and Kaspar Vereide [2,3]

1 Waterpower Laboratory, NTNU—Norwegian University of Science and Technology, Alfred Getz' vei 4, 7034 Trondheim, Norway; mads.ivarson@ntnu.no
2 Department of Civil and Environmental Engineering, NTNU—Norwegian University of Science and Technology, 7031 Trondheim, Norway; kaspar.vereide@ntnu.no
3 Sira-Kvina kraftselskap, Stronda 12, 4440 Tonstad, Norway
* Correspondence: chirag.trivedi@ntnu.no

Abstract: In order to increase the lifespan of hydraulic turbines in hydropower plants, it is necessary to minimize damages caused by sediment erosion. One solution is to reduce the amount of sediments by improving the design of sand trap. In the present work, the effects on sand trap efficiency by installing v-shaped rake structures for flow distribution and rib structures for sediment trapping is investigated numerically using the SAS–SST turbulence model. The v-shaped rake structures are located in the diffuser near the inlet of the sand trap, while the ribs cover a section of the bed in the downstream end. Three-dimensional models of the sand trap in Tonstad hydropower plant are created. The present study showed that integrating rib type structure can reduce the total weight of sediments escaping the sand trap by 24.5%, which leads to an improved sand trap efficiency. Consequently, the head loss in the sand trap is increased by 1.8%. By additionally including the v-shaped rakes, the total weight of sediments escaping the sand trap is instead increased by 48.5%, thus worsening the sand trap efficiency. This increases head loss by 12.7%. The results also show that turbulent flow commencing at the sand trap diffuser prevents the downstream settling of sediments with a diameter of less than one millimeter. The hydraulic representation of the numerical model is validated by comparison with particle image velocimetry measurements of the flow field from scale experiments and ADCP measurements from the prototype. The tested rib design has not previously been installed in a hydropower plant, and can be recommended. The tested v-shaped rakes have been installed in existing hydropower plants, but this practice should be reconsidered.

Keywords: hydropower; CFD; sand trap; sediment transport; particle; multiphase

Citation: Ivarson, M.M.; Trivedi, C.; Vereide, K. Investigations of Rake and Rib Structures in Sand Traps to Prevent Sediment Transport in Hydropower Plants. *Energies* **2021**, *14*, 3882. https://doi.org/10.3390/en14133882

Academic Editors: John M. Cimbala and Bryan J. Lewis

Received: 1 June 2021
Accepted: 23 June 2021
Published: 28 June 2021

1. Introduction

Sediment handling and erosion in hydropower plants have been long-standing engineering challenges. Hydropower plants are often upgraded and refurbished to improve performance and plant capacity. Sediment handling in hydropower plants can be done through catchment manipulation, dam and intake design, tunnel lining, sand traps, and turbine design. The motivation of the work is the upgrading and refurbishment of existing hydropower plants. After upgrading their installed capacity, several large Norwegian hydropower plants experienced operational problems associated with sediments entering the penstock and causing erosion to the turbine [1]. Moreover, a higher variability in power demand in recent years has led to some plants converting from base-load to peak-load production. This results in variable discharge through the tunnels, which stirs up sediments from the tunnel bed in unlined tunnels with remaining rock material on the invert, which is typical in Norwegian hydropower plants. Currently, the power plant does not have a system to indicate how much sediment actually passes through the sand trap and is transported through the turbine. To the authors experience, this is the typical situation for hydropower sand traps. For comparison, the records of removed sediments since 2015 are in the range from 150 m^3 to 15 m^3 per year, with a decreasing trend. Sediments can cause damage to the turbine through abrasion on the turbine surfaces. Larger particles may also cause larger dents on the turbines. Such damage results in reduced turbine efficiency and reduced structural integrity of the turbine components. The damage may, in turn, trigger increased cavitation that accelerates the degradation [2,3].

Tonstad hydropower plant, located in the mountains of southwestern Norway, is currently experiencing such challenges. In 1988, the plant upgraded its capacity from 640 to 960 MW by installing a new 320 MW Francis turbine. An additional penstock, surge tank, and sand trap were built. However, the unlined headrace tunnel was left untouched. It was expected that the higher discharge required to run all three turbines would lead to an increased amount of sand and rocks to be flushed into the turbines. However, the actual amount of sediments being transported with the flow was surprisingly large. One of the reasons for the increased amount of sediment transport is the poor design of the sand trap, which was designed in the 1960s and is now underperforming.

Figure 1 shows the layout of a typical high-head hydropower plant. A sand trap is a section of the water way, typically located immediately upstream of the penstock. This allows the headrace tunnel, with the remaining stone and sand material after its construction, to be unlined. Sand traps are designed to reduce flow velocity, which allows sediments to settle easier. The velocity is typically reduced by 30 to 50% [1]. This is done by increasing the cross-sectional area of the tunnel. The main challenge is poor performance of sand trap, which is intended to trap sediments before they reach the penstock. A proposal to improve sand trap performance is to cover sections of the bed with concrete ribs. This has been shown in physical experiments to create a low-velocity zone beneath the ribs, protecting sediments on the bed from being stirred up, while also allowing sediments to fall through the gaps between ribs [4,5]. Another proposal is to install rows of v-shaped rakes in the diffuser at the sand trap inlet. Increased levels of turbulence, such as those induced by these v-shaped rakes, has been shown to increase sediment settling speeds for certain particle sizes in numerical studies by Maxey in 1987 [6] and Wang and Maxey in 1993 [7]. This was later confirmed in physical experiments by Aliseda et al. in 2002 [8]. In these studies, sediments were found to settle in the peripheries of local vortex structures, which is coupled to a sweeping of sediments in directions normal to the flow.

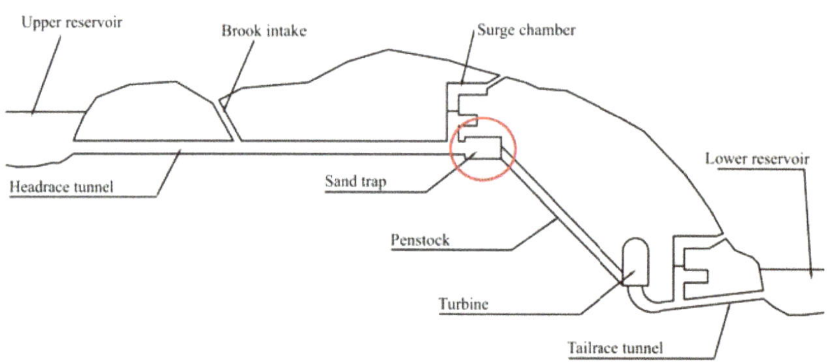

Figure 1. Layout of a typical high-head hydropower plant. The sand trap is marked with red color.

Several scientific studies have been conducted on the topic of sediment transport in hydropower sand traps. Olsen and Skoglund modelled the flow of water and sediment in a three-dimensional sand trap geometry using the $k-\epsilon$ turbulence model [9]. After including modifications to the turbulence model, both the flow field solution and sediment concentration calculations were in agreement with experimental procedures. Kjellesvig and Olsen modelled the bed changes in a sand trap using the transient convection–diffusion equation for sediment concentration and an adaptive grid adjusting for changes in the bed [10]. Large amounts of sediments could be seen being moved through the geometry in the simulations. The results compared well to physical model tests. Bråtveit and Olsen used 3D computational fluid dynamics (CFD) simulations to calibrate horizontal acoustic Doppler current profilers (H-ADCP) in the Tonstad sand trap [11]. The study found that the 3D CFD simulations could accurately calibrate the H-ADCP while also assessing the flow conditions at the locations of installation. Almeland et al. computed water flow in a model of the Tonstad sand trap using different versions of the $k-\epsilon$ turbulence model [12]. Depending on the discretisation scheme, grid resolution, and turbulence model, the computations showed the flow field to follow both the left side, right side, and centre of the diffuser. Field measurements showed that the main current followed the centre of the diffuser.

The present work is part of an ongoing sand trap research project. In previous work by Richter et al., it was found that implementing ribs just upstream of the penstock increased sand trap efficiency dramatically, as sediments were trapped in the low-velocity zone underneath the ribs [4,5]. Havrevoll et al. performed PIV analyses on the flow around ribs in the sand trap to investigate sediment settling characteristics. They showed that ribs successfully separate the flow field [13]. Daxnerova performed experiments on a physical scale model of the sand trap at NTNU to determine the effects of installing various calming flow structure designs in the diffuser [14]. The best performing design from Daxnerova's research, a v-shaped rake type structure, will be further studied in this work.

The objective of the present work is to assess the changes in sand trap efficiency when including ribs in the downstream end of the sand trap. The effects on the turbulence dissipation and sediment trajectories by including v-shaped rakes in the diffuser will be investigated. The work aims to reproduce results from experiments on physical scale models of the sand trap in order to gain further confidence in the experimental results.

2. Materials and Methods

2.1. Sediment Transport Theory

When sediments at the tunnel bed are subjected to shear stresses higher than what is required to loosen them from the bed load, the sediments will be suspended and transported with the flow. Sediment transport can be divided into two main categories, depending on the shear velocity to settling velocity ratio of the particles, $u^* > w$ [15]. These categories are suspended load and bed load. Suspended load consists of finer particles that have low inertia and settling velocities due to their low mass. They are therefore transported further by the flow before settling compared with bed load sediments. Bed load usually consists of rocks and larger grains of sand that are too heavy to be suspended in water for longer durations. They are instead transported by sliding, saltating, or rolling along the sediment bed, as illustrated in Figure 2.

In order to determine if the upgrades improve sediment settling, it is necessary to measure the sand trap efficiency. The most straightforward method to compute the efficiency is a sediment-mass-based approach. Here, the ratio of the total mass of sediments injected at the inlet and escaped through the outlet is found. The efficiency can then be calculated as in Equation (1)

$$\eta = 1 - \frac{\Phi_{s,out}}{\Phi_{s,in}}, \tag{1}$$

where η is the sand trap efficiency and Φ represents the mass of sediment. Time-integrated particle mass flow reports created at the end of simulations will be used to find the inflow and outflow of sediments and then calculate sand trap efficiencies.

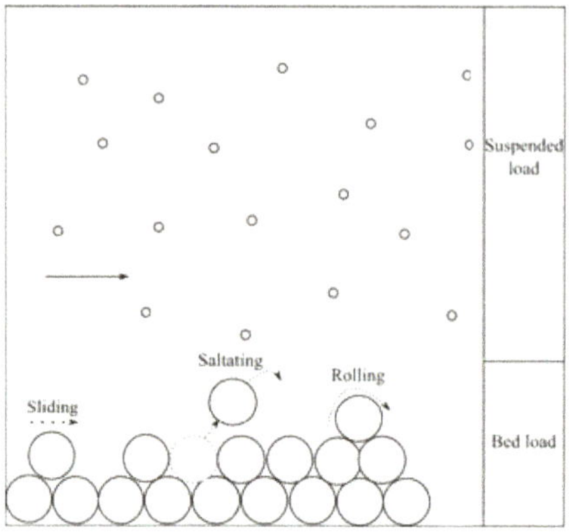

Figure 2. Types of sediment transport.

2.2. Head Loss

Installing new structures in the waterways could increase energy losses in the flow. This is mainly due to increased friction. This will affect the total head loss and, ultimately, the performance of the power plant. The head loss should therefore be minimised. Losses due to friction are determined by using the Darcy–Weisbach friction factor, seen in Equation (2).

$$h_f = f \frac{L \times v^2}{D \times 2g}, \tag{2}$$

where h_f (m) is the head loss, f is the dimensionless Darcy friction factor, L (m) is the length of the pipe, v (m/s) is the mean velocity of the flow, D (m) is the diameter of the tunnel, and g (m/s^2) is the gravitational acceleration. To ensure correct head loss calculations, it is important to find the correct Darcy friction factor. Using a Moody diagram, the friction factor can be found by determining the Reynolds number and relative roughness of the flow situation. In the present work, the Darcy–Weisbach equation form based on the pressure drop (Equation (3)) will be used to find the head loss caused by the upgrades.

$$\Delta h_L = \frac{\Delta p}{\rho g}, \tag{3}$$

where Δh_L is the head loss, Δp (Pa) is the pressure difference between two points, and ρ (kg/m^3) is the fluid density. The head loss in each of the geometries will be measured in post-processing. The head loss in the base case will be subtracted from each of the other cases, where the difference is the increased head loss caused by the upgrades.

2.3. Turbulence Modelling

To model the turbulent flow behaviour in the present work, the scale-adaptive simulation shear stress transport (SAS–SST) turbulence model was used. The model was found to perform better than the conventional RANS formulations in similar cases during the underlying project work. It introduces the Von Karman length scale into the turbulence scale equation to adapt to different turbulence structure sizes, while using base RANS equations in areas where the flow behaves more similar to steady state [16]. The SAS–SST are as follows:

$$\frac{\partial k}{\partial t} + \frac{\partial u_j k}{\partial x_j} = P_k - \beta^* \omega k + \frac{\partial}{\partial x_j}\left[(\nu + \frac{\nu_t}{\sigma_k})\frac{\partial k}{\partial x_j}\right] \tag{4}$$

$$\frac{\partial \omega}{\partial t} + \frac{\partial}{\partial x_j}(\overline{u}_j \omega) = \frac{\partial}{\partial x_j}\left[(\nu + \frac{\nu_t}{\sigma_\omega})\frac{\partial \omega}{\partial x_j}\right] - \beta\omega^2 + C_\omega + \alpha S^2(1 + P_{SAS}) \quad (5)$$

$$\nu_t \propto \frac{k}{\omega}, \quad P_{SAS} = \overline{\zeta}_2 \kappa \frac{L}{L_{vK,3D}}, \quad L_{vK,3D} = \kappa \frac{S}{U''}. \quad (6)$$

In these equations, S and U'' are generic first- and second-velocity derivatives, respectively. The SAS–SST model builds on the SST k-omega model by implementing an extra production term in the ω-equation, P_{SAS}. This term is attuned to transient fluctuations in the flow. In regions with a fine mesh where the flow is becoming unsteady, $L_{vK,3D}$ is reduced, increasing the production term. This will result in a large ω and, therefore, reduced k and ν_t values. In this way, the unsteadiness is not dampened, but is instead included as a part of the turbulence that is being resolved, leading to greater accuracy. A reduction of the turbulent viscosity dissipation occurs, which makes the momentum equations interpret the flow as transient rather than steady [17].

2.4. Computational Setup

The Tonstad power plant sand trap is 184 m-long, and the main tunnel has a cross-sectional area of around 9.9 m × 11 m (around 119 m^2). A picture of the prototype sand trap is shown in Figure 3. A 3D model of the sand trap is developed using engineering CAD software, see Figure 4. It was created from engineering drawings provided by the Sira-Kvina power company—the owner of the plant—and consists of a rectangular inlet section, a diffuser, a long tunnel section with a gentle slope, and a weir in the invert where the tunnel transitions into the penstock. This model is used as a base to add upgrades to in two other models. The additional models were created to test the effects of the upgrades, where one model includes only ribs and another includes both ribs and v-shaped rakes.

Figure 3. Picture of a prototype sand trap tunnel at hydropower plant; courtesy of Sira-Kvina kraftselskap.

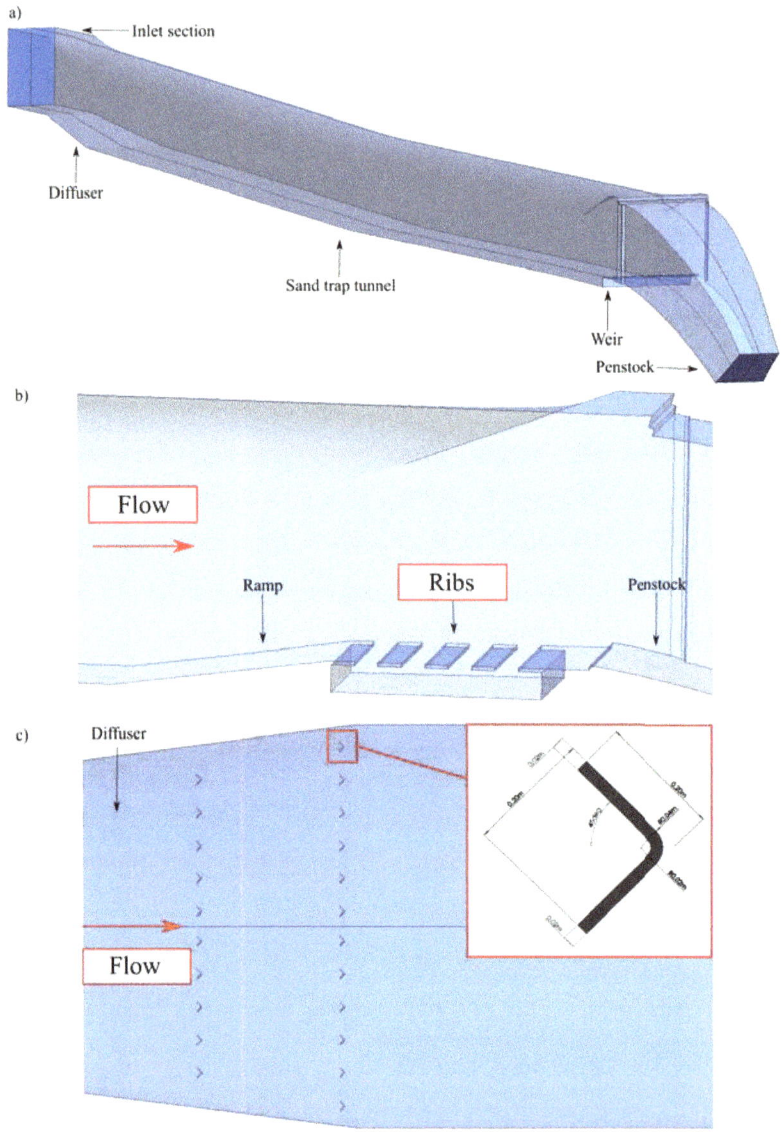

Figure 4. (**a**) Sand trap without upgrades. The distance from the end of the diffuser to the weir is 184 m. The diffuser is 16 m-long. The sand trap tunnel has a cross-sectional area of 119 m^2. (**b**) Ribs. The ribs are placed just upstream of the penstock, and are 1 m-wide and spaced 1 m-apart. The ramp has an 8% inclination. (**c**) Top view of the v-shaped rakes in the diffuser. The v-shaped rakes measure 6 m in height. Distances tip-to-tip between rakes are 1 m and 0.8 m for the upstream and downstream row, respectively. Zoomed part shows rake dimensions.

The ribs are placed just upstream of the penstock, in combination with the weir. This setup was tested in experiments [4,5]. The purpose of the ribs is to allow bed load sediments to fall between the ribs, while water is minimally affected. There are five ribs in total. The length of the ribs and the gap between ribs both measure 1 m. A ramp is placed upstream of the ribs to raise them above the bed and create space for sediments to accumulate. Further, in this way, excavating into the tunnel floor is not necessary. In the work by Richter et al., the ramp was also found to protect sediments that had fallen between the ribs from being resuspended and flushed into the penstock.

The rake structure is made up of two rows of v-shaped rakes, with the tip of the rakes facing downstream. The rakes measure 6 m in height. Distances between rakes are 1 m and 0.8 m for the upstream and downstream row, respectively. The purpose of the rakes is to even the flow of the jet from the inlet and enhance the diffuser effect of slowing down the flow. The distance from the inlet to the rakes is 25 m, which is approximately 3 times the diameter. To ensure that stable and developed flow reaches the rakes, a distance of 5–10 times the inlet diameter is necessary. The fact that the flow reaching the rakes may not be fully developed in the simulations should be taken into consideration when analysing the results.

The three models of the sand trap are meshed similarly. In all three models, the diffuser and the invert are both given tetrahedral mesh structures, while the simpler geometries of the inlet section, tunnel, and penstock have structured hexahedral meshes. The number of cells in each mesh is >23 × 10^6. Inflation layers are added along the tunnel walls so that global $y^+ < 30$. This ensures accurate representation of flow conditions in the boundary layers. The mesh surrounding the ribs and the rakes are refined further to a maximum cell size of 0.01 m. Average element sizes in the diffuser, tunnel, and rib section are 1.05 mm, 1.39 mm, and 0.43 mm, respectively.

The chosen advection scheme is the "High Resolution" scheme available in CFX. This scheme is second-order accurate in smooth regions and reduces its order of accuracy in regions of high gradients, where unboundedness is a factor. A first-order-accurate scheme is generally more robust and reaches convergence criteria faster than a second-order-accurate scheme. However, this comes at the cost of higher numerical diffusion, resulting in a less-accurate solution. The transient scheme used is the Second-Order Backward Euler scheme.

Steady state multiphase simulations are run to create initial conditions for the transient multiphase simulations. The total simulated time for the transient simulations is set to allow sediments enough time to either reach the bed or exit through the outlet. The simulations use a discharge of 80 m^3/s, which is the discharge when the power plant is operating at design conditions. The inlet velocity boundary condition represents this mass flow rate. The wall roughness is 10^{-3} m in the numerical model. An overview of solution parameters and boundary conditions is presented in Table 1.

Table 1. Selected parameters and boundary conditions for unsteady simulations.

Parameters	Description
Mesh type	Structured hexahedral and unstructured tetrahedral.
Model scale	1:1 to prototype.
Analysis type	Transient: total time of 1000 s, time step of 0.5 s.
Fluid	Incompressible Newtonian fluid, water density and viscosity at 10 C
Boundary conditions	Inlet: uniform velocity of 1.14 m/s. Outlet: total pressure of 0 Pa. Symmetry plane along center line.
Wall roughness	0.001 m.
Multiphase settings	Multiphase model: Discrete phase model/Particle transport solid. Fluid pair model: One-way coupled. Sediment mass flow rate: 1000 kg/s. Sediment particle diameter: $D_{particle} \sim$ N (0.75 mm, 0.25 mm)
Solver controls	Advection scheme: High Resolution. Transient scheme: Second-Order Backward Euler.
Turbulence model	SAS–SST.
Convergence control	Maximum residuals for pressure, mass-momentum and turbulent parameters < 10^{-4}. Coefficient loops: 2–3 iterations.

Multiphase flow was implemented by enabling the particle transport solid model in CFX, also known as the discrete phase model (DPM). DPM uses the Eulerian–Lagrangian multiphase model to track the paths of individual particles as they travel through the domain. It is well-suited for situations such as in the present work, where the volume fraction of the solid phase is low. The one-way coupled fluid pair model was chosen to solve the fluid–particle interactions. This model is computationally cheaper than the two-way coupled model. As the sediment phase's effect on the fluid phase is negligible in this case, the one-way coupled model was found to produce satisfactory results. In the transient multiphase simulations, sediments were injected with uniform distribution over the inlet during the first 100 s. The mass flow of sediments was set to 1000 kg/s, leading to a total mass of sediments injected into the sand trap of 10^5 kg. The sediments were tracked as they travelled through the model, and the sand trap efficiency was given by the time-integrated mass flow report at the outlet by the end of the simulation. The sediment diameters have a normal distribution with a mean of 0.75 mm and a standard deviation of 0.25 mm.

By also ensuring that the solution is mesh independent, the accuracy of the results are further improved. Mesh independence is achieved when a defining value of the simulation no longer changes significantly. The purpose of a mesh-independence study is to minimise the discretisation error resulted from approximating the geometry during the meshing stage [18]. The model with both rakes and ribs was used for the mesh-independence study. Three different mesh qualities were used. Following the procedure described by Celik I. B. et al. [19], an estimation of the discretisation error was obtained. In the calculations, N is the number of cells; h is the representative cell size; r is the refinement factor; ϕ is the pressure drop from inlet to outlet; p is the apparent order of the method; ϕ_{ext} are the extrapolated values of ϕ; e_a and e_{ext} are the approximate and extrapolated relative errors, respectively; and $\mathrm{GCI}_{\mathrm{fine}}$ is the fine-grid convergence index. The mesh discretisation error was found to be 1.4%. The results of the calculations are presented in Table 2. The fine mesh with 26.9 million cells is considered for further analysis and will be used to conduct the final numerical studies.

Table 2. Parameters of the mesh-independence study.

Parameter	Value
N_1, N_2, N_3	$26.9 \times 10^6, 9.7 \times 10^6, 4.3 \times 10^6$
h_1, h_2, h_3	0.08, 0.11, 0.14
r_{21}	1.4
r_{32}	1.3
ϕ_1	1838
ϕ_2	1860
ϕ_3	1897
p	2.1
ϕ_{ext}^{21}	1.8×10^3
e_a^{21}	0.0068%
e_{ext}^{21}	0.0458%
GCI_{fine}^{21}	1.4%

The simulation results are verified by ensuring that residuals reach a satisfactory convergence criteria and that there is a stable mass flow through the domain. This signifies that the solution is computationally correct. The hydraulic representation of the numerical model will be validated by comparison with PIV measurements on a physical scale model of the sand trap and to ADCP measurements from the prototype sand trap [4,20]. As the flow state is highly stochastic, the velocity distributions will never be identical. However, it is useful to make sure that the jet in the diffuser observed in the PIV measurements also exists in the simulated flow. It is also important to ensure that the simulated velocities are of reasonable magnitudes.

3. Results and Discussions

The results will mainly focus on how the velocity, vorticity, turbulence, and settling patterns vary between the three distinct models, as these are the key factors that affect sand trap efficiency in this case.

3.1. Sand Trap Efficiency

The sand trap efficiencies of the different models were obtained by creating a time-integrated particle mass flow report. This measures the total sediment mass entering the domain through the inlet and exiting through the outlet. The total time of 1000 s was found to be sufficient for all suspended sediments to exit the domain while remaining sediments are travelling along the bed. In the models with ribs implemented, sediments travelling along the bed are observed to pour into the gaps between ribs. At the end of simulation, not all sediments travelling along the bed have reached the ribs. However, a precedent is set by the sediments that do reach the ribs. These are all seen to pour into the the ribs instead of passing over them, indicating that this will also be the case for the remaining sediments. Due to limited computational resources, running the simulations for longer was not feasible. Running the simulations until all sediments are either completely settled or out of the domain could affect the sand trap efficiencies.

Using the model without upgrades as the base line, the mass of sediments exiting through the outlet is reduced by 24.5% from 2.5×10^3 to 1.9×10^3 kg by including the ribs. This indicates that the ribs are more effective at capturing and trapping bed load sediments than only the weir. By also adding the v-shaped rakes, the amount of sediments escaping the sand trap is increased by 48.5% from 2.5×10^3 to 3.7×10^3 kg compared with the model without upgrades. The amount of larger sediments trapped can be assumed to be similar before and after including the rakes, as larger sediments cannot be seen to escape the sand trap in neither particle track plots. The lack of performance from the model with rakes included can therefore be attributed to the turbulent vortices preventing smaller sediments to settle. This reduces sand trap efficiency. The sand trap efficiency of the different models are listed in Table 3.

Table 3. Sand trap efficiencies.

Model	Sediments Injected	Sediments Exited	Sand Trap Efficiency
No upgrades	10^5 kg	2.5×10^3 kg	97.5%
Ribs	10^5 kg	1.9×10^3 kg	98.1%
V-shaped rakes and ribs	10^5 kg	3.7×10^3 kg	96.3%

In all simulations, the divide between suspended load and bed load appears to be around 1 mm in diameter. The majority of sediments that remain suspended until escaping the sand trap are smaller than 1 mm, while the larger sediments travel along the bed by sliding, saltating, or rolling. Models for simulating sediment resuspension were not included in the simulations. The bed load sediments are therefore not observed to be resuspended. It has previously been shown that for classically dimensioned sand traps, sediment resuspension mostly occurs for sediments with grain sizes smaller than 2×10^{-4} m [21]. This is below the range of grain sizes used in the present work. Further analyses could be done with the resuspension models included and with smaller grain sizes to investigate the rate of sediment resuspension from the bed load. To further improve the sand trap efficiency, a flow calmer in the shape of horizontal bars, as suggested by Richter, should be tested. Instead of acting as a bluff body and inducing turbulence, this flow calmer could break up turbulence structures and improve settling characteristics for smaller sediments. Smoothing the transition between inlet and diffuser by reducing the inclination of the slope was suggested as an option to further increase the settling of smaller sediments [5]. However, it was found that this solution does not significantly improve the jet flow behaviour in the diffuser, thus not improving the settling of small sediments.

Looking at particle track plots of the different models, it appears that particles that reach the tunnel bed before passing over the ribs will indeed fall between them. This confirms the discovery from experiments by Richter et al. [4,5]. Sediments that remain suspended when passing the ribs will generally escape the sand trap. The amount of suspended sediments vary depending on if rakes are included in the diffuser or not. In the results where the rakes are not included, the suspended sediments are gathered closer towards the bottom of the tunnel. When rakes are included, the suspended sediments are of greater numbers and are more dispersed over the tunnel cross-section. This is believed to be caused by the turbulence from the rakes.

3.2. Head Losses

The head loss, Δh_L, of the different models was calculated from the steady-state simulations using the pressure-drop-based Darcy–Weisbach equation in Equation (3). The pressure difference, Δp, is calculated between the inlet and outlet faces of the models. Using the head loss in the model with no upgrades as a base value, the increased head loss caused by the upgrades was calculated by finding the difference in head loss between each of the upgraded models and the base value.

As presented in Table 4, the head loss caused by including just the ribs is 0.003 m, equating to an increase of 1.8% for the whole sand trap. Combined with the better sand trap efficiency of this model, this speaks for the value of including ribs in the sand trap. The model with both ribs and rakes included sees an increase in head loss of 12.7% compared to the model with no upgrades. The large head loss and the relatively poor sand trap efficiency of this model make it possible to conclude that other types of improvements to the sand trap should be pursued instead.

Table 4. Head loss, Δh_L, is calculated using Equation (3). Increased head loss is found by comparison with the model with no upgrades.

Model	Δh_L	Increased Head Loss
No upgrades	0.166 m	-
Ribs	0.169 m	0.003 m (+1.8%)
V-shaped rakes and ribs	0.187 m	0.021 m (+12.7%)

3.3. No Upgrades

The model with no upgrades represents the sand trap as it stands today, with a diffuser near the inlet and a weir just upstream of the penstock. The simulation results on the model with no upgrades give a baseline with which results from the other models can be compared. In addition, the results on this model will be compared to PIV and ADCP measurements for validation [4,5]. The velocity contour plot in Figure 5 shows the separation occurring at the entrance of the diffuser and the jet forming above it. Large circulation zones develop both in the horizontal and vertical planes at the entrance of the diffuser. These phenomena were obtained in both PIV results and in other experimental results, thus validating the simulations in this work [13,20]. Field measurements by Almeland et al. showed that the main current follows the centre of the diffuser, which can also be seen in the present results [12].

The turbulence, which develops from the separation in the diffuser, is seen to propagate through the sand trap, see Figure 6. The turbulence appears to dissipate as the flow reaches the halfway point before increasing as it crosses the weir and enters the penstock. The slow dissipation of turbulence may be due to the relative smoothness of the tunnel walls. Increasing the wall roughness to closer resemble the rough unlined tunnel walls in the prototype would affect the simulation results. One possibility is that turbulence would dissipate faster because of the increased energy losses. This would lead to improved sediment settling characteristics in the downstream end of the sand trap. Another possibility is that the rough walls may introduce even higher turbulence, disturbing sediment settling.

Figure 5. Velocity distributions in the sand trap with no upgrades included at t = 1000 s. A high-velocity jet above vortices caused by flow separation can be seen in the diffuser. Further downstream, the velocity is more evenly distributed.

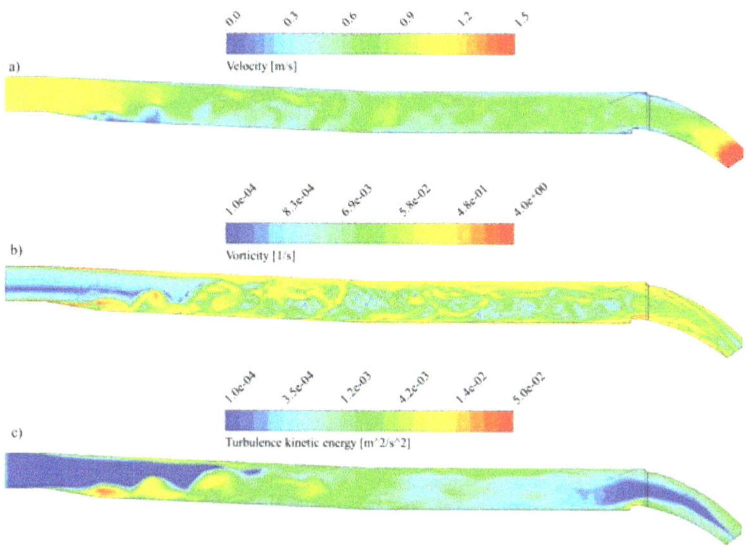

Figure 6. Sand trap without upgrades, symmetry plane at t = 1000 s. (**a**) Velocity contour. Flow separation occurs in the diffuser, which causes a higher flow velocity in the upper part of the diffuser. Separation is also seen to occur at the weir. (**b**) Vorticity contour. Flow separation in the diffuser and at the weir causes vortex generation. (**c**) Turbulence kinetic energy contour. Turbulence propagating from the diffuser starts to dissipate before reaching the penstock.

3.4. Sand Trap with Ribs

The flow behaviour upstream of the ribs remains identical to the model without upgrades. Large vortex structures propagate from the diffuser, where flow separation occurs. The separation of the flow field around the ribs is presented in Figure 7. This results in low velocities in the space below the ribs, which improves sediment settling. It can also be seen that inflow occurs at the last rib. This causes circulation in the downstream end of the space below the ribs. Sediments begin to settle in the upstream end, and will therefore be less affected by this circulation. However, this could change as the space fills up with sediments. A turbulent boundary layer forms over the ribs from separation at the ramp. This will be beneficial for the settling of bed load sediments under the ribs, as these will slow down when entering the boundary layer. The chance of the sediments falling through the gaps is therefore increased. Flow into the penstock is more turbulent as a consequence of the turbulent boundary layer.

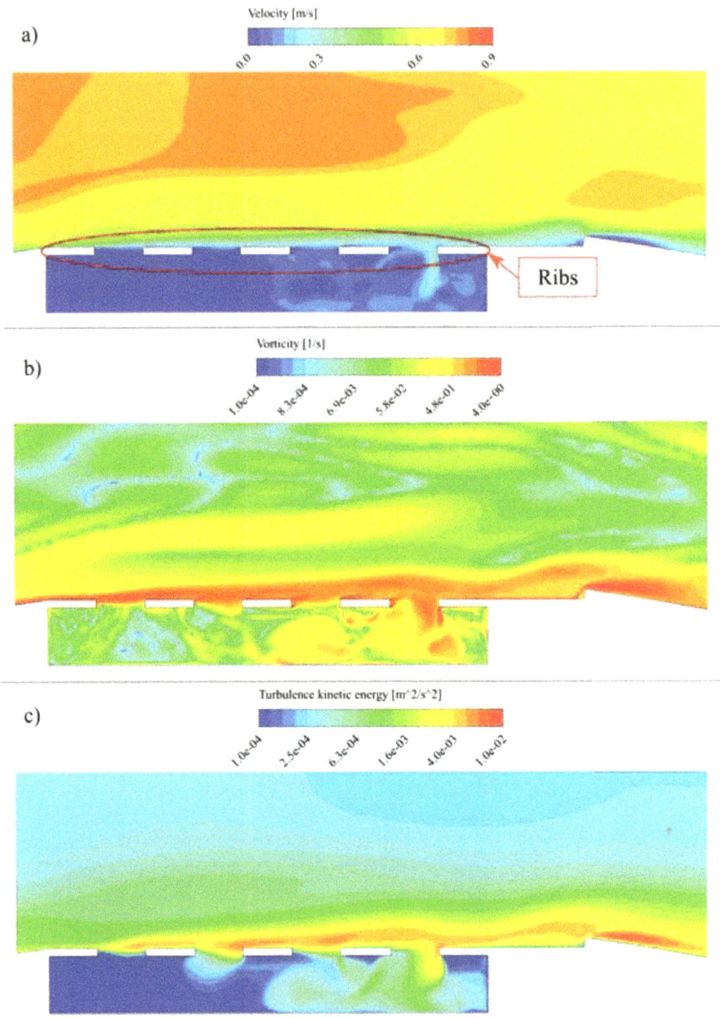

Figure 7. Extended view of ribs in the symmetry plane at t = 1000 s. (**a**) Velocity contour. Separation of the velocity field is visible. Low velocities below ribs increase the chances of sediment settling. Low-velocity inflow occurs between the last two ribs. This causes circulation in the downstream end below the ribs. Sediments will begin to settle in the upstream end, and will therefore be less affected by the circulation. However, this could change as the space fills up with sediments. (**b**) Vorticity contour. (**c**) Turbulence kinetic energy contour. A turbulent boundary layer forms over the ribs due to separation from the ramp. This will be beneficial for the settling of bed load sediments, as these will slow down when entering the boundary layer. Flow into the penstock is more turbulent as a consequence.

3.5. Sand Trap with Rakes and Ribs

It was hypothesised that the higher turbulence induced by the rakes would increase settling speed for larger sediment sizes. By studying the particle tracks in the simulation results, it can be seen from Figure 8 that sediments with a diameter larger than 1 mm tend to settle earlier in the sand trap compared with the geometries without the rakes in the diffuser. However, it can be observed that sediments smaller than 1 mm tend to remain suspended for longer when rakes are included. These smaller sediments have the potential to cause erosion damage on the turbine blades, and it is therefore desired to prevent these from escaping the sand trap.

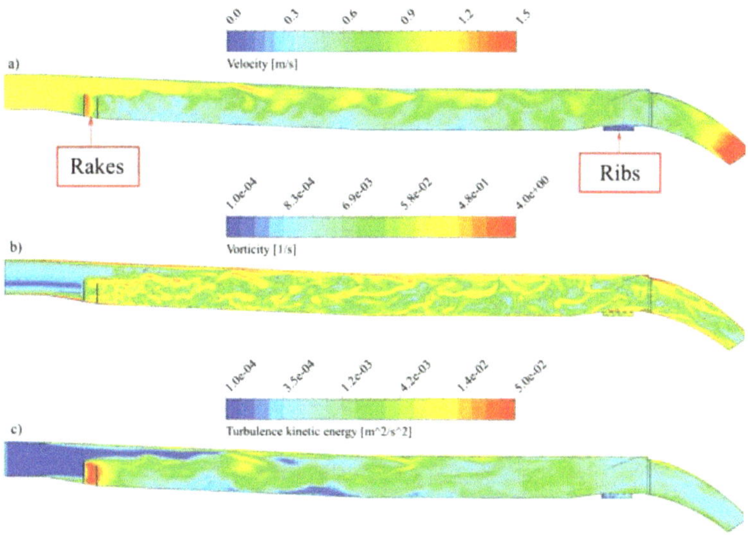

Figure 8. Sand trap with ribs and v-shaped rakes, symmetry plane at t = 1000 s. (**a**) Velocity contour. Flow over the rakes is accelerated, while flow going through the rakes slows down and becomes turbulent. (**b**) Vorticity contour. High vorticity appears immediately downstream of rakes and remains throughout the sand trap. (**c**) Turbulence kinetic energy contour. Rakes induce higher levels of turbulence than can be seen in models without rakes. Turbulence has not dissipated before the flow exits the sand trap

The large circulation zones, which also occur in the models without rakes, can be seen clearly in Figure 9. Sediments are seen to become trapped in these circulation zones in particle track plots. The flow is separated as it passes the rakes, where flow going over is accelerated, while flow going through decelerates and becomes turbulent. Vorticity and turbulence are induced by the vortex shedding at the rakes. The highest levels of turbulence are observed between the two rows of rakes. A large turbulent wake is established downstream of the rakes and remains until the outlet. In the present work, this has been shown to decrease sand trap efficiency. The flow downstream of the diffuser when rakes are included is seen to be more turbulent than when rakes are omitted. As the turbulence does not dissipate before exiting the sand trap, this causes more turbulent flow to enter the penstock.

If the height of the rakes was to be increased so that they reach the crown of the tunnel, it could affect the settling characteristics in multiple ways. One possibility is that increasing the height of the rakes would cause an earlier onset of turbulence and vorticity, which again carries small diameter sediments further. From the results in the present work, it is believed that this would lead to an increase in head loss and a decrease in sand trap efficiency. Another possibility is that the flow would no longer be divided into high- and low-velocity zones downstream of the rakes. Instead, a general reduction in absolute flow velocity would occur. This could mean that the flow becomes more uniform, which might be beneficial for sand trap efficiency. In both cases, increasing the flow obstructing area is likely to increase head loss.

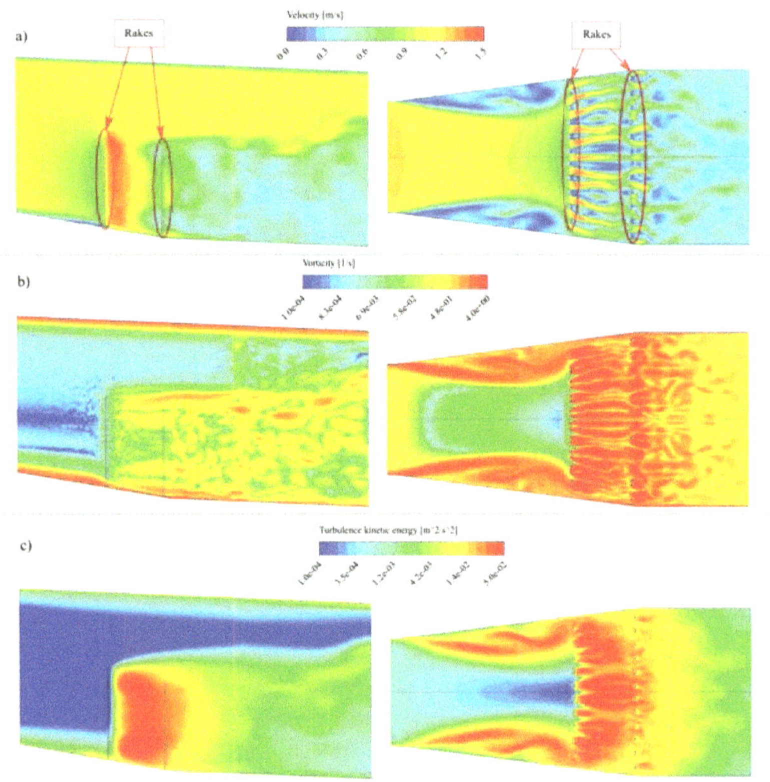

Figure 9. Extended view of v-shaped rakes in symmetry and horizontal planes at t = 1000 s. (**a**) Velocity contours. The large circulation zones, which can also be seen in the models without rakes, appear at the entrance of the diffuser. These zones are observed trapping sediments. Flow is separated going past the rakes. Flow going over is accelerated, while flow going through is decelerated. Velocity is highest between rakes. (**b**) Vorticity contours. Vorticity and turbulence is induced by the vortex shedding at the rakes. (**c**) Turbulence kinetic energy contours. The highest levels of turbulence are observed between the two rows of rakes. A large turbulent wake is established downstream of the rakes, which has been shown to decrease sand trap efficiency in the present work.

3.6. Consequences for Sand Trap Design

Based on the analysis results it is found that the tested rib design improves the trap efficiency and can be recommended for new sand traps and retrofitting of existing sand traps. The tested design was recently developed and presented in Richter et al. [5] and has not previously been installed in a hydropower plant. The tested v-shaped rakes were found to decrease the trap efficiency and, in addition, have a larger negative effect on the head loss. Such v-shaped rakes are installed in several existing hydropower plants, but this practice should be reconsidered. However, it is noted that only one design of the v-shaped rakes were tested and other designs may prove to have a positive effect on the trap efficiency.

4. Conclusions

It is necessary to minimise sediment erosion to increase the turbine's lifespan. Three-dimensional models based on the Tonstad power plant sand trap were created. Versions of the model include various upgrades to determine their effect on sediment settling. The numerical domain was discretised by a combination of hexahedral and tetrahedral mesh. Steady-state and transient multiphase simulations were performed on the models, using water and sand with a variable grain size. The objective was to investigate how installing rake- or rib-type structures affect particle sand trap efficiency and head loss.

By investigating the results, it was found that the sand trap with ribs at the outlet reduces the total weight of sediments exiting the sand trap by 24.5%, while increasing the head loss by around 1.8%. Installing rakes in the diffuser, although showing signs of increasing settling speed for larger sediments, was found to increase the total weight of sediments leaving the sand trap by 48.5%. This led to a reduced sand trap efficiency. In addition, the rakes caused an increased head loss of 12.7%. It is shown that in all models, sediments escaping the sand trap have a diameter smaller than one millimetre. These findings are supported by physical scale experiments on the sand trap [4,5,13]. The results show that installing ribs at the outlet of the sand trap will reduce sediment transport to the turbine and increase sand trap efficiency, thus prolonging turbine lifespan at the Tonstad power plant.

The main novelties from this work are the analysis results for the tested design of ribs and v-shaped rakes. The tested rib design has previously not been installed in a hydropower plant, and can be recommended based on the results in this work. The tested v-shaped rakes have been installed in existing hydropower plants previously, but this practice should be reconsidered.

For further work, running two-dimensional simulations along the centre line of the sand trap with the Large Eddy Simulation turbulence model could give a more accurate representation of the turbulence and sediment settling in the plane. Additional variants of the rakes should be tested to verify that they still have an adverse effect. The effects of extending the rakes to reach the crown of the tunnel so that the whole flow area is covered should also be investigated. Additionally, a model allowing for sediment resuspension should be explored to better represent sediments bouncing on or being resuspended from the bed. Further, experimental measurements of the inlet velocity profile are needed to create realistic and accurate inlet boundary conditions.

Author Contributions: Conceptualization, M.M.I., C.T. and K.V. methodology, M.M.I.; software, M.M.I.; formal analysis, M.M.I.; investigation, M.M.I.; resources, C.T.; data curation, M.M.I.; writing original draft preparation, M.M.I.; writing—review and editing, C.T. and K.V.; visualization, M.M.I.; supervision, C.T. and K.V.; project administration, C.T. All authors have read and agreed to the published version of the manuscript.

Funding: No funding is received for this work.

Institutional Review Board Statement: Not applicable.

Informed Consent Statement: Not applicable.

Conflicts of Interest: The authors declare no conflict interest.

1. References

1. Richter, W.; Vereide, K.; Zenz, G. Upgrading of a Norwegian pressurized sand trap combined with an open air surge tank. *Geomech. Tunn.* **2017**, *10*, 620–624. [CrossRef]
2. Truscott, G.F. A literature survey on abrasive wear in hydraulic machinery. *Wear* **1972**, *20*, 29–50. [CrossRef]
3. Thapa, B.S.; Thapa, B.; Dahlhaug, O.G. Current research in hydraulic turbines for handling sediments. *Energy* **2012**, *47*, 62–69. [CrossRef]
4. Vereide, K.; Richter, W.; Havrevoll, O.H.; Betete, K.; Shrestha, U.; Navaratnam, U.; Lia, L.; Mauko, G. *Flexible Sandtraps: Final Report (HydroCen Report)*; Technical Report; Norwegian Research Centre for Hydropower Technology: Trondheim, Norway, 2021; Volume 20.

5. Richter, W.; Mauko, G.; Zenz, G. *Hydraulic Investigation, Numerical and Physical Model Test, Flexible Sand Trap 2.0*; Technical Report; Graz University of Technology: Graz, Austria, 2020.
6. Maxey, M.R. The gravitational settling of aerosol particles in homogeneous turbulence and random flow fields. *J. Fluid Mech.* **1987**, *174*, 441–465. [CrossRef]
7. Wang, L.P.; Maxey, M.R. Settling velocity and concentration distribution of heavy particles in homogeneous isotropic turbulence. *J. Fluid Mech.* **1993**, *256*, 27–68. [CrossRef]
8. Aliseda, A.; Cartellier, A.; Hainaux, F.; Lasheras, J.C. Effect of preferential concentration on the settling velocity of heavy particles in homogeneous isotropic turbulence. *J. Fluid Mech.* **2002**, *468*, 77–105. [CrossRef]
9. Olsen, N.R.B.; Skoglund, M. Three-dimensional numerical modeling of water and sediment flow in a sand trap. *J. Hydraul. Res.* **1994**, *32*, 833–844. [CrossRef]
10. Olsen, N.R.B.; Kjellesvig, H.M. Three-dimensional numerical modelling of bed changes in a sand trap. *J. Hydraul. Res.* **1999**, *37*, 189–198. [CrossRef]
11. Bråtveit, K.; Olsen, N.R.B. Calibration of Horizontal Acoustic Doppler Current profilers by three dimensional CFD simulations. *Eng. Appl. Comput. Fluid Mech.* **2015**, *9*, 41–49. [CrossRef]
12. Almeland, S.K.; Olsen, N.R.B.; Bråveit, K.; Aryal, P.R. Multiple solutions of the Navier-Stokes equations computing water flow in sand traps. *Eng. Appl. Comput. Fluid Mech.* **2019**, *13*, 199–219. [CrossRef]
13. Havrevoll, O.H.; Vereide, K.; Rüther, N.; Lia, L. *PIV Experiments on Ribs in the Tonstad Rock Trap Model*; NTNU Reportno. B1-2021-2; Norwegian University of Science and Technology: Trondheim, Norway, 2021; p. 18.
14. Daxnerová, J. Hydraulic Scale Modelling of Flow Calming Structures for Hydropower Plants. Master's Thesis, NTNU, Trondheim, Norway, 2019.
15. Fergus, T.; Hoseth, K.A.; Sæterbø, E. *Vassdragshåndboka*, 1st ed.; Fagbokforlaget: Trondheim, Norway, 2010.
16. Menter, F.R.; Egorov, Y. The Scale-Adaptive Simulation Method for Unsteady Turbulent Flow Predictions. Part 1: Theory and Model Description. *Flow Turbul. Combust.* **2010**, *85*, 113–138. [CrossRef]
17. Davidson, L. The SAS Model: A Turbulence Model with Controlled Modelled dissipation. In Proceedings of the 20th Nordic Seminar on Computational Mechanics, Göteborg, Sweden, 20–23 November 2007.
18. Roache, P.J. Perspective: A Method for Uniform Reporting of Grid Refinement Studies. *J. Fluids Eng.* **1994**, *116*, 405–413. [CrossRef]
19. Celik, I.B.; Ghia, U.; Roache, P.; Freitas, C. Procedure for Estimation and Reporting of Uncertainty Due to Discretization in CFD Applications. *J. Fluids Eng.* **2008**, *130*. [CrossRef]
20. Brevik, O. *3D Numerisk Modellering av Deler av Vannvegen til Tonstad Kraftverk*; Institutt for vann-og miljøteknikk: Trondheim, Norway, 2013; Volume 139.
21. Ortmanns, C. Entsander von Wasserkraftanlagen. Ph.D. Thesis, ETH Zurich, Zurich, Switzerland, 2006.

energies

MDPI

Article

Flow Characteristics of Preliminary Shutdown and Startup Sequences for a Model Counter-Rotating Pump-Turbine

Jonathan Fahlbeck *, Håkan Nilsson and Saeed Salehi

Department of Mechanics and Maritime Sciences, Chalmers University of Technology, SE-412 96 Gothenburg, Sweden; hakan.nilsson@chalmers.se (H.N.); saeed.salehi@chalmers.se (S.S.)
* Correspondence: fahlbeck@chalmers.se

Abstract: Pumped Hydropower Storage (PHS) is the maturest and most economically viable technology for storing energy and regulating the electrical grid on a large scale. Due to the growing amount of intermittent renewable energy sources, the necessity of maintaining grid stability increases. Most PHS facilities today require a geographical topology with large differences in elevation. The ALPHEUS H2020 EU project has the aim to develop PHS for flat geographical topologies. The present study was concerned with the initial design of a low-head model counter-rotating pump-turbine. The machine was numerically analysed during the shutdown and startup sequences using computational fluid dynamics. The rotational speed of the individual runners was decreased from the design point to stand-still and increased back to the design point, in both pump and turbine modes. As the rotational speeds were close to zero, the flow field was chaotic, and a large flow separation occurred by the blades of the runners. Rapid load variations on the runner blades and reverse flow were encountered in pump mode as the machine lost the ability to produce head. The loads were less severe in the turbine mode sequence. Frequency analyses revealed that the blade passing frequencies and their linear combinations yielded the strongest pulsations in the system.

Keywords: hydropower; pumped hydro storage; low-head; counter-rotating; pump-turbine; transient sequences; shutdown; startup; OpenFOAM

Citation: Fahlbeck, J.; Nilsson, H.; Salehi, S. Flow Characteristics of Preliminary Shutdown and Startup Sequences for a Model Counter-Rotating Pump-Turbine. *Energies* **2021**, *14*, 3593. https://doi.org/10.3390/en14123593

Academic Editor: Andrea De Pascale

Received: 30 April 2021
Accepted: 14 June 2021
Published: 16 June 2021

1. Introduction

The demand to control and regulate the electrical grid to provide grid stability is ever increasing. This is a consequence of the fact that intermittent renewable energy sources, e.g., wind and solar, are on the rise and will continue to increase in the coming decades [1]. Pumped Hydro Storage (PHS) is the maturest and most cost-efficient solution to store energy and thus provide grid stability [2–4]. The oldest PHS power plants date back to the late 19th Century [5]. In the year 2019, PHS had a capacity of 158 GW worldwide [3]. Traditionally, PHS facilities require very specific site locations to make them economically feasible. According to Deane et al. [6], the head of the PHS facility is the most essential criterion, and a higher head is preferable. Most research is thus focused on PHS intended for higher heads. The ALPHEUS H2020 EU project ("Augmenting Grid Stability Through Low Head Pumped Hydro Energy Utilization and Storage") aims to contribute with economically viable PHS solutions for low- to ultra-low-heads [7,8]. The main goals of the ALPHEUS project are to achieve a reversible pump-turbine with a power of 10 MW, a round-trip efficiency of 70–80%, and heads in the region of 2–20 m. The ALPHEUS project focuses on three pump-turbine designs, a shaft-driven Counter-Rotating Pump-Turbine (CRPT), a rim-driven CRPT, and a positive displacement configuration. In the present paper, a model scale of an initial design of the shaft-driven CRPT is considered.

Wintucky and Stewart [9] concluded already in the 1950s that a counter-rotating turbine may have 2–4% higher overall efficiency compared to a single-rotor configuration. The concept of a counter-rotating propeller configuration is commonly associated with marine or aeroplane propulsion systems [10]. In recent years, the counter-rotating propeller configuration has been investigated as a reversible hydropower pump-turbine unit. Furukawa et al. [11] stated in 2007 that a CRPT, compared to a single rotor, can have a smaller shroud diameter, a lower rotational speed, and a wider range of operation at high efficiency with an individual speed control of the two runners. In a multi-objective optimisation of a CRPT, Kim et al. [12] showed that a hydraulic efficiency of close to 80% in turbine mode and above 85% in pump mode could be achieved, for a wide range of operating conditions. A CRPT has been analysed in pump mode with unsteady Computational Fluid Dynamics (CFD) simulations and validated with experimental test data by Momosaki et al. [13]. It was concluded that unsteady computations are required to accurately predict the performance of the CRPT, even at the design point.

Recent studies of Francis-like pump-turbines have shown that it is possible to numerically predict the flow characteristics during transient sequences such as load-rejection and mode-switching [14–17]. The current study investigated the shutdown and startup sequences for a model-scale CRPT, in both pump and turbine modes. In conventional Francis-like pump-turbines, guide vanes are used to direct and control the flow [18]. In contrast to the various Francis-like pump-turbines, the shaft-driven CPRT in the ALPHEUS project has no guide vanes. It is the head and individual rotational speeds of the runners that determine the point of operation. The transient sequences include a decrease of the runner rotational speeds, from the design point to stand-still. The rotational speeds are later increased from stand-still back to the design point.

2. Design and Operating Conditions

Figure 1 shows the blade geometry of the analysed CRPT, as well as the computational domain. Conventional Francis-like pump-turbines utilise guide vanes to produce angular momentum in turbine mode and to reduce angular momentum in pump mode, thus ensuring high efficiency in both modes. This is not the case for the CRPT as the basic principle of the CRPT is that the upstream runner is designed to have an axial inflow, while the counter-rotating downstream runner makes use of the angular momentum leaving the upstream runner, generating a close to axial flow downstream the runners at the best efficiency point. Figure 1a shows the blade geometries of Runner 1 (red) and Runner 2 (blue). Runner 1 worked as the upstream runner in pump mode and downstream in turbine mode. In this study, the CRPT was at model scale, since the numerical results are later to be validated against experimental test data. Figure 1b shows the full computational domain, the locations of velocity probing lines, and pressure probe P4. The computational domain included the two runners, the hub with support struts, and contraction/expansion sections before/after the machine to focus the kinetic energy. Some parts of the straight pipes were included in the computational domain to keep the boundary conditions at some distance. The total length was 12.8-times the diameter of the runners. In pump mode, the flow was from left to the right, and in turbine mode, it was from right to left.

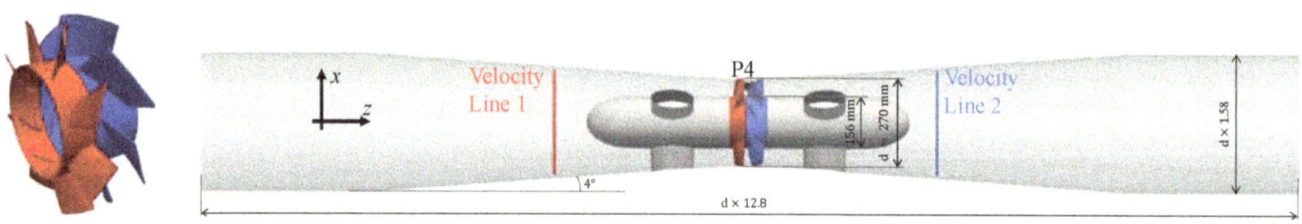

(a) (b)

Figure 1. Blade geometry and computational domain. (**a**) Runner 1 (red) and Runner 2 (blue). (**b**) Computational domain with the coordinate system, geometrical dimensions, location of pressure probe P4, and velocity lines 1 and 2. In pump mode, the flow is from left to right (positive z) and, in turbine mode, from right to left (negative z).

The blade geometries, shown in Figure 1a, were designed by Advanced Design Technology (ADT) Ltd., as a part of the ALPHEUS H2020 EU project [19]. Runner 1 had eight blades, while Runner 2 consisted of seven blades. The two runners rotated in opposite directions from one another. No guide vanes were utilised, as the upstream runner was designed to have an axial inflow. The runner hub and shroud diameters were 156 mm and 270 mm, respectively. The machine was operated by individually controlling the rotational speed of each runner. At the design point in pump mode, Runner 1 rotated at 1423 rpm and Runner 2 at 1307 rpm. The corresponding rotational speeds in turbine mode were 832 rpm and 749 rpm. The engineering quantities at the design point are summarised in Table 1. The net head and power were higher in pump mode since the machine must overcome the hydraulic losses at the test facility. In turbine mode, those losses were subtracted, yielding a lower net head and power. The hydraulic efficiency of the machine was roughly the same in both modes. The head was defined by the total pressure drop of the computational domains of the runners, coloured red and blue in Figure 1b. The efficiency is calculated in pump and turbine modes as:

$$\eta_{\text{Pump}} = \frac{\rho g H Q}{P}, \tag{1}$$

$$\eta_{\text{Turbine}} = \frac{P}{\rho g H Q}. \tag{2}$$

Here, ρ is the fluid density, g is the gravity acceleration, H is the head, and P is the power. The power was in this work defined as $P = T_{\text{R1}}\Omega_{\text{R1}} + T_{\text{R2}}\Omega_{\text{R2}}$, where T is the torque and Ω is the rotational speed in rad/s of the runners. The subscripts R1 and R2 are for Runner 1 and Runner 2, respectively.

Table 1. Engineering quantities at the design point in pump and turbine modes.

	Pump	Turbine	Unit
Head	15.3	9.2	m
Power	56.0	20.1	kW
Discharge	333.5	252.9	L/s
Efficiency	89.2	88.2	%

3. Methods

The present CRPT was previously evaluated in detail at the design point in both pump and turbine mode, with both steady-state and unsteady numerical simulations [20]. In the present study, shutdown and startup sequences were simulated and analysed using CFD. The numerical simulations were carried out with the OpenFOAM-v1912 open-source CFD software [21,22]. As previously mentioned, the runners rotated at individual speeds. For a given net head, it was the combination of rotational speeds of the runners that controlled the operating point of the machine. Two full shutdown and startup sequences were carried out in the present work, one for each mode. The rotational speeds were first decreased, from the design point to stand-still, and later increased back to the design point. The evaluated transient sequences are shown in Figure 2. The rotational speeds are presented as functions of time for the two modes. At time $t = 0$ s, the flow was fully developed at the design point, based on the previous simulations. The shutdown sequences started at time $t_{0,1} = 0.2$ s in both modes. The time from the design point to complete stand-still was 2.6 s and 4.5 s in turbine and pump mode, respectively. The runners were at a stand-still between times $t_{\text{end},1}$ and $t_{0,2}$. The startup sequences started at $t_{0,2}$ and continued to $t_{\text{end},2}$. The entire simulation time was 8.0 s in turbine mode and 11.3 s in pump mode. The transient sequence required less time in turbine mode as the rotational speeds were lower at the design point compared to in pump mode. The rotational speeds were decreased and increased symmetrically by a sinusoidal function as:

$$n_r = \frac{n_{r,\text{DP}}}{2}\left[1 + \sin\left(\pi \frac{t - t_0}{t_{\text{end}} - t_0} \pm \frac{\pi}{2}\right)\right]. \tag{3}$$

Here, n is the rotational speed in rpm, n_{DP} is the rotational speed at the design point, t is the time, t_0 is the start time of the transient, and t_{end} is the end time of the transient. Index r is 1 or 2 for the corresponding runner. The $\pm$ sign is positive when decreasing the rotational speed and negative when increasing the speed. The sinusoidal shape to change the rotational speed of the runners was chosen since a sinusoidal function allows a smooth transition between two constant rotational speeds.

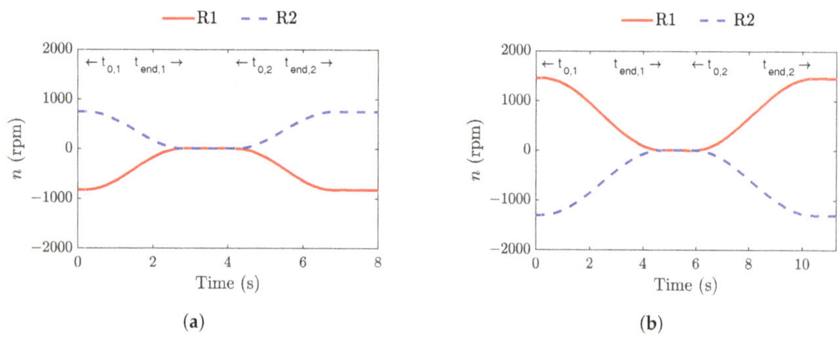

Figure 2. Rotational speeds of the runners as a function of time during the transient sequences. Here, R1 and R2 denote Runner 1 and Runner 2, respectively, $t_{0,i}$ and $t_{end,i}$ correspond to the start respective end time of the transient operation. Index i is 1 for shutdown and 2 for startup. (**a**) Turbine mode: $t_{0,1} = 0.2$ s, $t_{end,1} = 2.8$ s, $t_{0,2} = 4.2$ s, $t_{end,2} = 6.8$ s. (**b**) Pump mode: $t_{0,1} = 0.2$ s, $t_{end,1} = 4.7$ s, $t_{0,2} = 5.8$ s, $t_{end,2} = 10.3$ s.

3.1. Governing Equations and Numerical Schemes

The incompressible Navier–Stokes equations were discretised and solved for a computational mesh using the finite volume method [23,24]. The k-ω SST-SAS eddy viscosity model was employed to take into account the effects of unresolved turbulence. The SAS modifications to the standard k-ω SST turbulence model allow for a decrease in the turbulent viscosity where the turbulence may be partially or fully resolved [25]. This turbulence model has recently been successfully used in a number of studies concerning transient operations in a hydropower context [14,15,26]. The simulations were carried out on the full 3D computational domain shown in Figure 1b. Each runner domain rotated individually with a solid body rotation, employing an Arbitrary Mesh Interface (`AMI`) for the transfer of fluxes at the sliding interfaces [27,28].

The convection terms of the momentum equations were discretised using the second-order accurate Linear-Upwind Stabilised Transport (`LUST`) scheme [29]. This is a blended scheme, utilising a 75% central differencing scheme (`linear`), for accuracy, and a 25% second-order upwind scheme (`linearUpwind`), for stability [30]. The first-order `upwind` scheme was used to discretise the convection terms in the transport equations for turbulent kinetic energy (k) and specific turbulence dissipation rate (ω). The `linear` scheme with the Gauss theorem was employed for the gradient schemes of all variables. The implicit, second-order accurate scheme (`backward`) was used for temporal discretisation. A constant time step of $\Delta t = 5 \times 10^{-5}$ s was used. At the design point, in pump mode, the time step corresponded to a Courant number of less than 2.2 in 90% of the numerical domain, and the average Courant number was less than 0.065. The maximum runner rotation was 0.44° per time step.

The pressure–velocity coupling was handled using the `PIMPLE` algorithm [30]. The algorithm allows the Courant number to exceed unity by combining the `PISO` and `SIMPLE` algorithms. The solver algorithm in the present work was configured with two inner `PISO` loops and a maximum of ten outer `SIMPLE` corrector loops. One non-orthogonal corrector step was used within each inner loop. The `PIMPLE` algorithm interrupts the outer corrector loops if convergence is reached within each time step. The solution was deemed as converged if the absolute tolerance was below 10^{-6} and 10^{-5} or the relative tolerance was below 0.1 and 0.01, for velocity and pressure, respectively. The absolute tolerance is the solver residual, and the relative tolerance is in relation to the initial solver residual. The solver algorithm always converged within three or four outer corrector loops.

3.2. Boundary Conditions

A total pressure boundary condition was used for the inlet, while a static pressure boundary condition was imposed on the outlet boundary. The flow rate was thus calculated as a part of the solution. A total-to-static pressure difference of 100 kPa and 134 kPa was used in turbine and pump mode, respectively. The total-to-static pressure difference over the computational domain ensured the net head of the CRPT at the design point, presented in Table 1. Note that the net head in the table was defined by the total pressure difference over just the CRPT, and the total-to-static pressure difference concerned the entire computational domain. In OpenFOAM-v1912, the `totalPressure` boundary condition was used for pressure at both the inlet and the outlet, together with the velocity boundary condition `pressureInletOutletVelocity`. This combination of boundary conditions has recently been used to accurately simulate a shutdown transient for a high-head Francis model turbine by Uppström et al. [31]. The `totalPressure` boundary condition corresponds to a static pressure formulation on the boundary as:

$$p_{\text{boundary},i} = p_0 - \frac{\rho|\vec{u}_i|^2}{2}, \text{ for inflow (at face } i\text{)}, \tag{4}$$

$$p_{\text{boundary},i} = p_0, \text{ for outflow (at face } i\text{)}. \tag{5}$$

Here, p_0 is the user-specified pressure (total for inflow and static for outflow, hence the minus in Equation (4)), ρ is the fluid density, and $\vec{u}$ is the velocity vector. Index i corresponds to face i on the boundary, meaning that the conditions are applied on a face-by-face basis. The boundary condition imposes a constant total pressure at the inlet faces and constant static pressure at the outlet faces. The `pressureInletOutletVelocity` boundary condition for the velocity applies a homogeneous Neumann (zero gradient) condition on the outlet faces. At the inlet faces, the velocity is obtained from the internal field in the face normal direction, as described by Fahlbeck [32]. It was essential to use a pressure–velocity boundary condition combination that allows for reverse flow, since the flow direction changes during the shutdown and startup operations in pump mode.

3.3. Computational Mesh

The computational mesh was divided into four regions, one for each runner and one for each of the contraction/expansion regions upstream and downstream the runners. The meshes of the runners were mostly block-structured, whereas the contraction/expansion regions consisted of 6 layers of prism boundary layer cells near the walls and unstructured tetrahedral core cells, as shown in Figure 3. The shroud tip clearance was 0.67 mm, and 8 layers of hexahedral and triangular prism cells were used in the tip clearance region. The mesh was finest at the runners, and gradually coarser further away from the runners. The surface mesh of one blade passage for the two runners is shown in Figure 3a, and the refined mesh regions close to the runners are shown in Figure 3b.

Table 2 presents a summary of the meshes for different regions. Note that the upstream and downstream regions utilised the same mesh. The total number of cells was 7.75 million, and the different regions contained roughly the same number of cells. The $y^+_{90\%}$ values in Table 2 denote the maximum y^+ values in 90% of the region. They were obtained in pump mode at a time step before the transient operations had commenced. The y^+ value was less than 50 in 90% of the computational domain. Wall functions that adapted to all $y+$ values were used. In OpenFOAM-v1912, the wall boundary conditions `kqRWallFunction` for turbulent kinetic energy (k), `omegaWallFunction` for specific dissipation rate (ω) and `nutkWallFunction` for turbulent viscosity were applied.

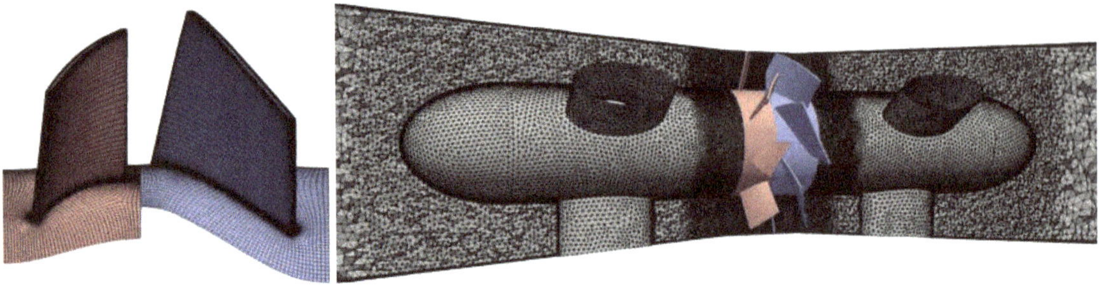

(a) (b)

Figure 3. Computational mesh regions, Runner 1 (red) and Runner 2 (blue). (**a**) Surface mesh of one blade passage. (**b**) Zoomed-in view of the refined mesh regions and surfaces of the runners.

Table 2. Mesh details.

Region	Cells × 10^6	$y^+_{90\%}$
Runner 1	1.90	45
Runner 2	1.97	54
Upstream/downstream	2×1.94	46
Full domain	7.75	50

4. Results and Discussion

The outcome of the numerical simulations is presented and discussed in this section. The section is divided into two sub-sections, where first, the analysis in turbine mode is discussed, followed by the pump mode investigation. Recall the time and rotational speeds shown in Figure 2, as both are referred to in this section.

4.1. Turbine Mode Transient Sequence

The shutdown and startup sequences in turbine mode are analysed here. The numerically predicted flow rate is shown in Figure 4 as a function of time. The flow rate was initially 253 L/s, at the design point. The rotational speed of the runners started to decrease at $t_{0,1}$, and the flow rate followed the decreasing rotational speed. This was because with a lower rotational speed, the CRPT was not effectively extracting energy from the flow, and the losses that reduced the flow rate were increasing. As the runners were at a stand-still ($t_{\mathrm{end},1} < t < t_{0,2}$), the machine was simply a large obstacle in the flow path, and the flow rate was 132 L/s. This was almost half the flow rate of the design point. At $t_{0,2}$, the runner rotational speeds started to increase, and the flow rate increased accordingly. The change of flow rate occurred symmetrically in the turbine mode sequences with the change of the rotational speeds. This was not the case in pump mode, as discussed later.

With a decreasing rotational speed, unfavourable flow structures increased in scale and number, and the flow became massively separated after the CRPT, as shown by the vorticity magnitude in Figure 5. Note that the flow was from top to bottom, and that vorticity is defined as the curl of the velocity. At $t = 0$ s (Figure 5a), the flow was rather axial, both before and after the runners, and the vortex shedding of the downstream support-strut was captured. As the rotational speeds decreased, it can be seen in Figure 5a–d that the flow at the blades of Runner 1 (lower one) first started to separate, then the same happened also at Runner 2. The reduction in rotational speed of Runner 2 increased the swirl coming to Runner 1. At the same time, Runner 1 was also decreasing its rotational speed, which further deteriorated the relative flow angle at the leading edges of Runner 1's blades. This is why the flow behind Runner 1 separated much more than that of Runner 2. This suggests that Runner 1 should maintain its rotational speed in the initial phase, but at some point, it must reduce its speed to stand-still and thus go through such unfavourable conditions.

However, at that point, there would also be a valve that closed and further reduced the flow rate of the machine. Suggesting the optimum shutdown and startup sequences to minimise the damaging effects of the large flow separation structures remains to be investigated. At time $t = 4$ s, the runners were both at a stand-still, and a chaotic flow field had developed downstream the runners. This was mainly generated by a massive separation on Runner 1's blades, although also, the separation on Runner 2's blades increased significantly. As the rotational speed of the runners increased, the efficient tandem operation of the two runners became apparent again. This is shown by Figure 5d–g, as the unfavourable flow structures diminished during the startup sequence. As for the flow rate variation, shown in Figure 4, the variation in unfavourable flow structures (and thus losses) was very similar for the shutdown and startup sequences. However, the vortex shedding behind the downstream support-strut needed some additional time to develop after $t = 7$ s.

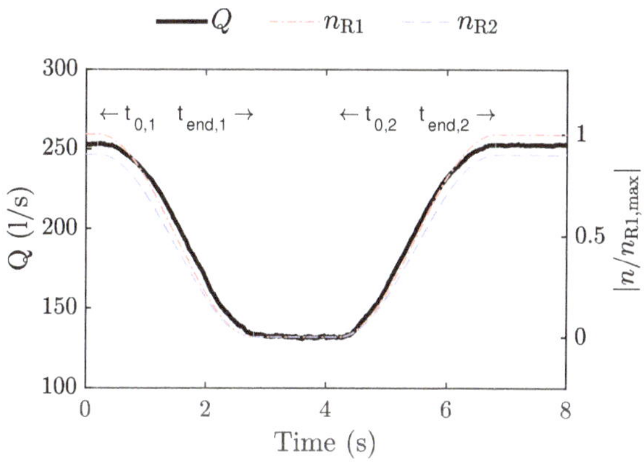

Figure 4. Flow rate (left axis) and runner rotational speeds (right axis) as a function of time during the turbine mode transients. The rotational speeds are normalised with the initial rotational speed of Runner 1. The annotated times correspond to the start/end times of the changes in runner rotational speeds (see Figure 2a).

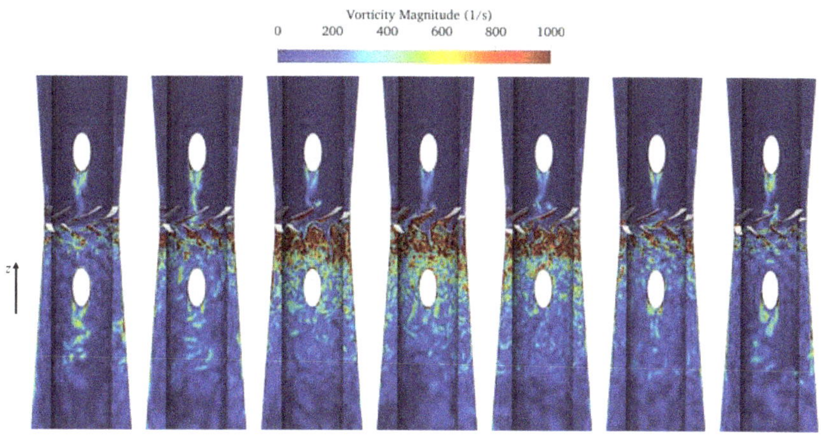

(**a**) $t = 0$ s. (**b**) $t = 1$ s. (**c**) $t = 2$ s. (**d**) $t = 4$ s. (**e**) $t = 5$ s. (**f**) $t = 6$ s. (**g**) $t = 7$ s.

Figure 5. Zoomed-in view of the vorticity magnitude at various time steps during the transient sequences in turbine mode. The results are displayed on a meridional plane and a cylinder (midpoint radius between hub and shroud) cutting the domain. The flow is from top to bottom.

The axial velocity at probing Line 1 (see the location in Figure 1b) is shown in Figure 6a as a contour over the x-coordinate and time. The flow profile was rather similar over the centreline at the earlier and later time steps, where the machine was working at the design point condition. There was not a distinct difference in the largest magnitude of the axial velocity at $t < 1$ and $t > 7$ s, for positive and negative x-coordinates. This was despite the fact that a support-strut was located between the runners and the line for negative x-coordinates. This means that the flow was not purely axial downstream the runners and that there was some swirl leaving the runners. The remaining swirl rotated the flow profile, causing a close to symmetric flow profile at the velocity line. As the runners were at a stand-still ($t_{end,1} < t < t_{0,2}$), the flow profile was far from symmetric. At $x > 0$ cm, close to zero and reverse flow were encountered at the velocity line. This was explained by the velocity vectors shown in Figure 6b. At the position of Line 1 (red), the flow rotated back towards the runners. This was while the flow was in the correct direction at the negative x-coordinates. The reason for the flow appearance at the velocity line was connected to the location of Line 1 in relation to the stand-still position of the runner blades. When the runners were at a stand-still, wakes downstream the runners developed depending on how the runners were oriented. In general, the flow was rather chaotic downstream the runners as the rotational speeds were small or zero. This was because the flow was massively separated at this point, and the resolved solution was expected to show randomness as the SAS model resolved part of the turbulent spectrum. The solver residuals were at this stage in the order of 10^{-6} for momentum, 10^{-5} for pressure, and 10^{-10} for continuity, which indicated a relatively small error at each time step.

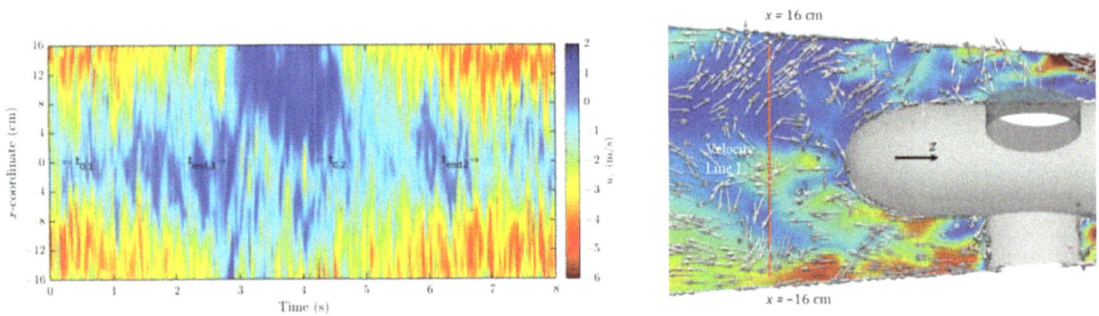

(**a**) (**b**)

Figure 6. Axial velocity contour and a zoom-in view of a meridional plane with velocity vectors during the transient sequence in turbine mode. (**a**,**b**) use the same colour bar and vertical axis. The axial velocity is here defined as positive in the negative z-direction (see Figure 1b). (**a**) Contour of axial velocity at Line 1 (downstream) during the turbine mode transients. (**b**) Axial velocity and velocity vectors at $t = 4$ s. Velocity Line 1 is shown in red.

Figure 7a,b shows the absolute axial force and absolute torque, respectively, during the turbine mode transients. The forces and torque acting on the runners were from integrated normal and shear stresses on the hub and blade surfaces of the runners. In the discussions below, the word *absolute* is dropped for brevity. The axial force and torque showed a smooth variation without any drastic peaks. As the rotational speeds decreased, the loads of Runner 1 increased and the loads of Runner 2 decreased. The axial force and torque was to a large extent a function of the head, or total pressure drop between the two reservoirs. In this study, the *reservoirs* were represented by the pressure boundary conditions at the inlet and the outlet. Thus, the summation of the loads of the two runners must always be rather constant in turbine mode. As the flow of Runner 1 started to separate, due to the intensified swirl coming from Runner 2, the loads increased for Runner 1. As a consequence of the load increase for Runner 1, Runner 2 experienced a decrease in loads. The axial force on the upstream runner was slightly higher when the runners were at a stand-still. This was because Runner 2 faced the upper reservoir, and Runner 1 was in the wake of Runner 2. The torque was higher for Runner 1 since Runner 2 directed the flow in an unfavourable direction towards Runner 1 at a stand-still. When the startup sequence was initiated, at $t_{0,2}$, the upstream runner experienced an axial load increase while the downstream runner experienced a load decrease. The torque followed the trends of the axial force during the transient sequences. The force variation in the transverse directions for Runner 1 is shown in Figure 7c as a function of time during the transient sequences in turbine mode. The transverse force fluctuations were rather calm until the rotational speed was low. Between $t = 2$ s to $t_{\text{end},1}$ and $t = 5$ s to $t_{0,2}$, large transverse force fluctuations were detected for Runner 1. They most likely arose from the flow separation at the blades of Runner 1, shown in Figure 5. The flow separation caused the formation of large chaotic flow structures at Runner 1's blades. This generated large force fluctuations in the transverse directions. The transverse forces for Runner 2 (not shown here) did not show such fluctuations when close to stand-still, since Runner 2 did not experience such massive flow separation as Runner 1.

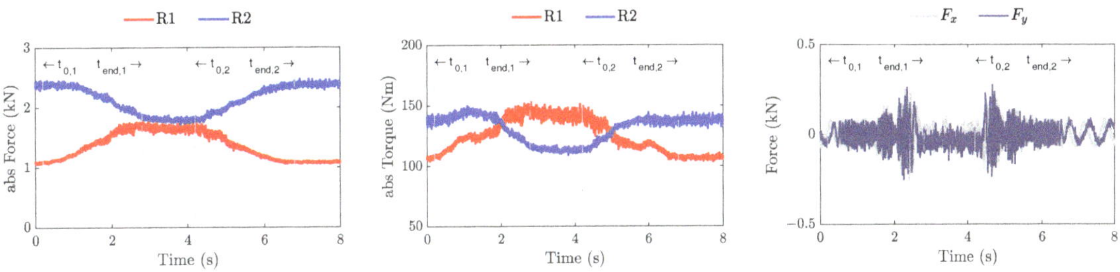

(**a**) Absolute axial force. (**b**) Absolute torque. (**c**) Runner 1 transverse forces.

Figure 7. Axial force and torque on the runners and transverse forces for Runner 1, as a function of time during the transient sequence in turbine mode. The *z*-component is the axial component.

Short-Time Fourier Transform (STFT) was performed to analyse the quantities in the frequency domain, due to the time-varying nature of the obtained signals. Frequency analysis was carried out on the fluctuating component of the pressure-probe P4, shown in Figure 1b, and the forces acting on the runner surfaces. The fluctuating component of a signal was calculated as:

$$\varphi' = \varphi - \bar{\varphi}. \tag{6}$$

Here, φ is the signal of an arbitrary quantity and $\bar{\varphi}$ is the instantaneous average of the signal. The instantaneous average was calculated with the Savitzky–Golay filter [33]. STFT is similar to performing a Fast Fourier Transform (FFT) in different time spans. The STFT uses a window averaging and the FFT average over the whole signal [34]. The window size should be chosen in a way to capture all the physical fluctuations. Here, a window size of 0.1 s was chosen to resolve all the high- and low-frequency oscillations. The frequency analysis was performed using the signal processing toolbox of MATLAB and visualised through a spectrogram. Pressure probe P4, located between the runners, was subjected to an STFT analysis in the turbine mode sequences. The fluctuating part of the pressure signal was obtained according to Equation (6). The static, average, and fluctuating pressures at P4 are shown in Figure 8a. It was evident from the pressure signal, p, that the instantaneous average, $\bar{p}$, was required in order to calculate the true fluctuating component, p'. This was because the pressure signal did not fluctuate around a fixed value as the rotational speeds were changed. Figure 8b shows an STFT, in a spectrogram, of the fluctuating part of the pressure at P4. The blade passing frequencies, $f_{\text{b,R1}}$ and $f_{\text{b,R2}}$, of the runners showed the highest power in the frequency domain. The second harmonic of the blade passing frequencies, f_{h2}, were also apparent for both runners, but evidently not as powerful as the main frequencies. The third harmonic, f_{h3}, was seen at an even weaker power than the second harmonic. Linear combinations of the blade passing frequencies, f_{l}, were also recognised through the transient. In a turbomachinery application, it is expected that the blade passing frequency is the dominating frequency close to the runners. This is because the runners produces and cuts strong wakes, resulting in pressure pulsations. The closer one is to the runners, the stronger the dynamics of the runners. The linear combination, f_1^- and f_1^+, of the blade passing frequencies were caused by runner–runner interaction, as the downstream runner cut the wakes of the upstream runner. Between the times $t \approx 2$ s and $t \approx 5$ s, a wider range of white noise was seen in the frequency domain. This was caused by the massive flow separation at the blades of runners, shown in Figure 5c–e. The white noise showed a peak in power at the same times as the oscillations for the transverse forces for Runner 1 increased (Figure 7c). The unfavourable oscillations noted as the rotational speed was small should be further analysed and minimised to increase the lifetime of the machine.

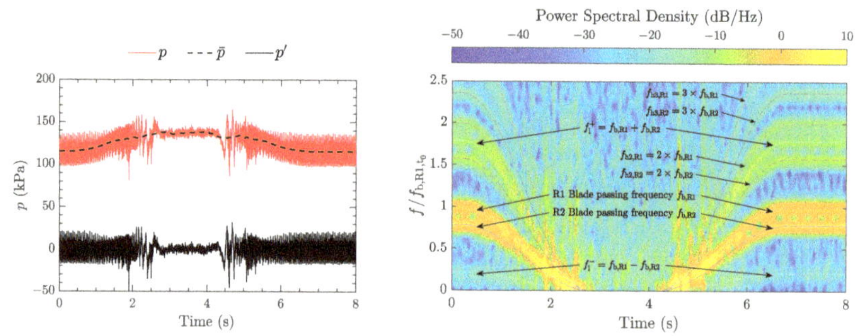

(**a**) P4 pressure signal as a function of time. (**b**) STFT of p' at probe P4.

Figure 8. Signal of pressure probe P4 and spectrogram of the STFT of the fluctuating pressure component, during the transient sequence in turbine mode. Here, p is the pressure signal, $\bar{p}$ is the instantaneous average, and p' is the fluctuating component. The frequency is normalised with, $f_{b,R1,t_0}$, the blade passing frequency of Runner 1 at t_0. Indices R1 and R2 denote Runners 1 and 2, respectively; b denotes the blade passing frequency; h2 is the second harmonic; h3 is the third harmonic; l is a linear combination.

4.2. Pump Mode Transient Sequence

In pump mode, the machine must produce sufficient head to overcome the height elevation and the hydraulic losses between the upper and lower reservoir. If the machine cannot produce enough head, the flow will change direction, and the CRPT no longer works as a pump, but as a mixer. During transient sequences in pump mode, the rotational speed of the runners changed according to Figure 2b. At the initial time step, the machine operated at the design point with a flow rate of 334 L/s and a net head of 15 m. Figure 9 shows how the flow rate varied as a function of time during the transient sequences in pump mode. The flow rate was here defined as positive in the preferred pump mode direction (positive z-direction) and negative as the flow was in reversed direction. The flow rate decelerated with the decreasing rotational speed of the runners. At t_{Q1}, roughly one third into the shutdown sequence, the pump lost the ability to produce enough head. After t_{Q1}, the flow accelerated in the reverse direction (identified here by a negative flow rate value). The rotational speeds of the runners were at this time 1031 rpm and -928 rpm for Runners 1 and 2, respectively. As the runners were at a stand-still, $t_{end,1} < t < t_{0,2}$, the flow rate was 153 L/s in the reverse direction. When increasing the rotational speeds, $t > t_{0,2}$, it was evident that there were hysteresis effects in the flow. Between $t_{0,2}$ and $t \approx 6.5$ s, the flow rate slowly decelerated. As the rotational speeds increased further, the machine started to build up some pressure. At $t > 9$ s, the flow rate was rapidly decelerated up to $t_{Q2} = 9.72$ s. At this point, the head of the machine balanced the head of the system as the flow rate was zero. The flow rate started to accelerate rapidly after t_{Q2}. Observe that the flow rate continued to accelerate after $t_{end,2}$, despite the fact that the runners had a constant speed after this point. This showed that it was not only the rotational speed of the runners that determined the flow rate. This was because it took time for the flow to adjust to the rotational speed, and it depended on if the flow was accelerated or decelerated to the point of operation.

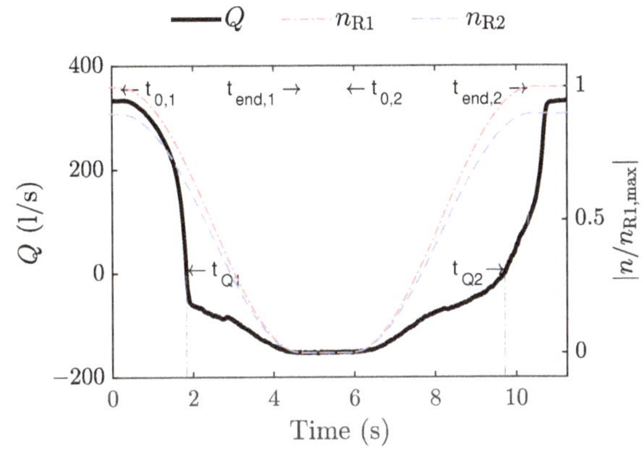

Figure 9. Flow rate (left axis) and runner rotational speeds (right axis) as a function of time during the pump mode transient. The flow rate is defined as positive in the preferred pump mode direction, positive z-direction (see Figure 1b). $t_{Q1} = 1.83$ s and $t_{Q2} = 9.72$ s indicate the points with zero flow rate. The rotational speeds are normalised with the initial rotational speed of Runner 1. The annotated times correspond to the start/end times of the changes in runner rotational speeds (see Figure 2b).

Flow structures during the transient sequences in pump mode are visualised in Figure 10 with the vorticity magnitude at a number of time steps. The preferred flow direction was from bottom to top, when the machine was functioning properly as a pump. The flow structures changed drastically in number and scale during the shutdown and startup sequences. At time $t = 0$ s, the CRPT operated at the design point. No large structures were apparent as the machine managed to maintain a close to axial flow downstream the runners. Vortex shedding from the support-struts was clearly visible at this stage. Already at $t = 1.5$ s, the machine was less efficient as a pump. This is shown in Figure 10b by the large angle of the vortices of the downstream support-strut. Between $t = 1.5$ and 2 s, the flow changed direction, and the machine was no longer able to pump the water in the preferred direction. At the intermediate time step, $t = 1.8$ s, the machine struggled to balance the flow, and large separation occurred by the blades of both runners. As time progressed (Figure 10d,e), the rotational speeds decreased further, and the large flow structures followed the reverse flow direction. At time $t = 6$ s, the runners were almost at a stand-still. The turbulent flow field was at this stage fully developed in the reverse flow direction, as shown in Figure 10f. The flow field was heavily separated by the blades of the runners at the stand-still position. The rotational speed now increased, and at $t = 9$ s, the runners struggled to build up sufficient head and force the flow in the correct pump mode direction. This is shown by the large flow structures on both sides of the runners in Figure 10g,h. By time $t = 10.5$ s, the machine pushed the flow in the preferred direction. This is shown in Figure 10g, as the flow structures now started to travel downstream the runners in the preferred direction. At the final time step, $t = 11.3$ s, most of the flow structures were flushed out of the domain, and the machine operated at the design point.

Figure 11 shows the axial velocity during the shutdown and startup sequences at Velocity Line 2 in pump mode. Velocity Line 2 is defined in Figure 1b, and it was located downstream the runners when the machine acted as a pump. Before the reverse flow was encountered, $t < t_{Q1}$, the peak axial velocity was found at a vertical position of $x \approx 10$ cm.

A similar, but smaller, peak was noted at $x \approx -10$ cm. The axial velocity was less at the negative x-coordinate due to the position of a support-strut in relation to Velocity Line 2. The axial velocity decelerated and accelerated quickly as the flow changed direction. The flow had a rather flat profile between $t \approx 3$ and 8 s. The velocity rapidly changed at $t \gtrsim 9$ s as the pump built up the head and pushed the flow in the correct direction. At the final time steps, $t > 10.5$ s, the familiar pattern from the initial time steps was present.

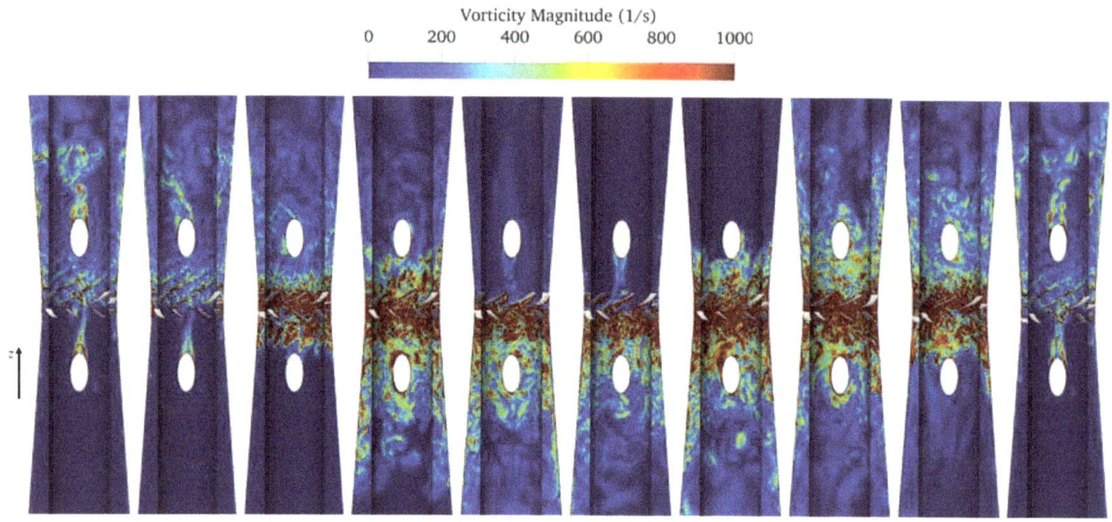

(**a**) t = 0 s. (**b**) t = 1.5 s. (**c**) t = 1.8 s. (**d**) t = 2 s. (**e**) t = 3 s. (**f**) t = 6 s. (**g**) t = 9 s. (**h**) t = 10 s. (**i**) t = 10.5 s. (**j**) t = 11.3 s.

Figure 10. Zoomed-in view of the vorticity magnitude at various time steps during the transient sequences in pump mode. The results are displayed on a meridional plane and a cylinder (midpoint radius between hub and shroud) cutting the domain. The preferred flow direction is from bottom to top in pump mode.

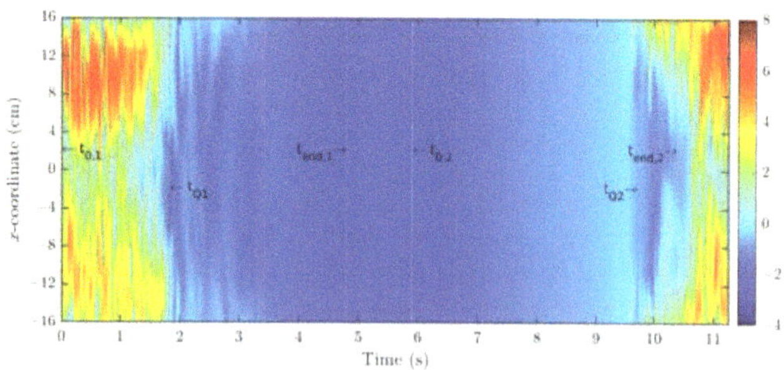

Figure 11. Contour of axial velocity at Line 2 (downstream in the preferred flow direction) during the pump mode transients.

Figure 12a,b show the absolute axial force and absolute torque as a function of time during the pump mode transients. In the following discussion, the word *absolute* is dropped, for brevity. The axial load on Runner 1, initially upstream, decreased with the rotational speed before the flow direction reversed. At the same time, the axial force on Runner 2, initially downstream, was rather constant. At the time of zero flow rate, t_{Q1}, a sudden decrease in both axial force and torque was encountered for Runner 2. As a consequence of this, Runner 1 had a rapid load increase. This means that Runner 2 stopped functioning as a pump before Runner 1 and that Runner 1 had to work against the head of the upper reservoir on its own to a larger extent. As the pressure drop over Runner 2 decreased, the loads on the runner followed. This produced a larger pressure drop, and loads, for Runner 1 as it struggled to produce sufficient head. As the rotational speeds continued to decrease, the axial load became similar for the runners. When the runners were at a stand-still, Runner 2 exhibited a slightly larger axial force than Runner 1. This was because Runner 2 was now upstream and facing the upper reservoir. Runner 1 was now downstream and located in the wakes of Runner 2. As the rotational speeds started to increase, the loads on Runner 1 followed, while the loads on Runner 2 decreased. When approaching the change in flow direction, t_{Q2}, the reverse phenomena was noted, and at t_{Q2}, the axial forces of the runners matched one another. After this point, Runner 1 experienced a large decrease in axial force and torque, and Runner 2 had a rapid increase, until the tipping-point at $t \approx 10.6$ s was reached. A plausible explanation for the load decrease for Runner 1 and the load increase for Runner 2 is as follows. Runner 2 managed to build up head before Runner 1, which means a large pressure drop and thus loads for Runner 2. This was because Runner 1 did not yet manage to produce sufficient pressure just upstream of Runner 2. After the tipping-point, the axial force and torque converged to their initial values. The rapid variations of the axial force and torque were most likely not a desirable feature, as large loads can damage or deteriorate the machine prematurely. By examining the forces in the transverse directions for Runner 1, shown in Figure 12c, large oscillations were noted between t_{Q1} and t_{Q2}. These force variations in the transverse direction most likely arose from the chaotic flow separation occurring by the blades of the runners. The transverse force fluctuations may have a negative impact on the lifetime of the machine. This is because large fluctuating forces are well connected to fatigue.

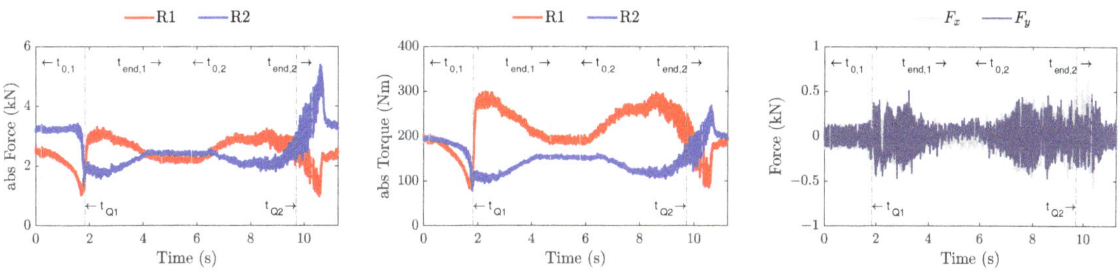

(**a**) Absolute axial force. (**b**) Absolute torque. (**c**) Runner 1 transverse forces.

Figure 12. Axial force and torque on the runners and transverse forces for Runner 1, as a function of time during the transient sequences in pump mode. The z-component is the axial component.

In Figure 13a,b, the STFT for pressure probe P4 and the transverse force $F_{R1,x}$ are presented. P4 was located between the runners, and $F_{R1,x}$ was the force in the x-direction on Runner 1. At the pressure probe P4, the blade passing frequencies of the runners, $f_{b,\,R1}$ and $f_{b,\,R2}$, showed a strong power through the transient sequences. At the initial and final time steps ($t_{Q1} > t > t_{Q2}$), the blade passing frequency of Runner 2, $f_{b,\,R2}$, showed a stronger power than the blade passing frequency of Runner 1, $f_{b,\,R1}$. As the flow rate changed direction ($t = t_{Q1}$), $f_{b,\,R1}$ exceeded $f_{b,\,R2}$ in intensity. The Runner 1 blade passing frequency was in fact stronger than that of Runner 2 until the flow was once again reversed to the preferred direction ($t = t_{Q2}$). Hence, it was found that the stagnation pressure of the downstream runner was always stronger than the wakes from the upstream runner at P4. This was because Runner 1 was initially upstream and Runner 2 downstream, so as the flow changed direction, the strongest blade passing frequency changed as well. The second harmonic frequencies of Runner 2, $f_{h2,R2}$, and Runner 1, $f_{h2,R1}$, showed the same behaviour as the main blade passing frequencies. The third harmonic frequencies, f_{h3}, and linear combinations, f_l^+ and f_l^-, of the blade passing frequencies were apparent at P4, but at a lower intensity. For the transverse force $F_{R1,x}$ (Figure 13b), the positive linear combination of the blade passing frequencies, f_l^+, of the runners showed the strongest power throughout the transients. The blade passing frequencies, $f_{\mathrm{b,R1}}$ and $f_{\mathrm{b,R2}}$, of the two runners were also clearly visible for $F_{\mathrm{R1},x}$, but not as strong as the positive linear combination. An explanation why it was the positive linear combination that had a decisive impact is as follows. The pulsations from each runner were caused by the wakes of each runner, and the linear combination, on the other hand, was caused by the cutting of the upstream runner wakes by the downstream runner. This phenomenon is shown at the design point in Figure 13c. Here, Runner 2 was downstream, and it cut the wakes of the upstream Runner 1. The wakes from the upstream runner were a low-pressure zone. It was cut by a high, stagnation, pressure from the leading edge of the downstream runner. This interaction impacted the pressure, and thus force, pulsations on the runners. The large pressure difference explained why the positive linear combination of the blade passing frequency had a strong impact on the transverse force component, $F_{\mathrm{R1},x}$. As stated by Lengani et al. [35], any linear combination of the blade passing frequencies can possibly be excited due to the runner–runner interaction of a counter-rotating machine. In addition to the positive linear combination, f_l^+, the negative linear combination, f_l^-, was also seen through the transient sequences. At the pressure probe (Figure 13a), the signal showed a significant amount of noise at the times related to a zero flow rate. This was because flow structures were formed and dissolved irregularly over a wide range of scales between the runners, as indicated by the vorticity in Figure 10. The pressure signal was rather smooth as the runners were at a stand-still. The STFT for the transverse force, shown in Figure 13b, revealed a wider range of frequencies and white noise in the signal. This was plausibly due to the flow separation occurring by the blades of the runner. The flow separation interacted with the transverse force over a wide range of scales, which explained the white noise. It was however the typical frequencies correlated to the rotation of the runners that had the strongest power.

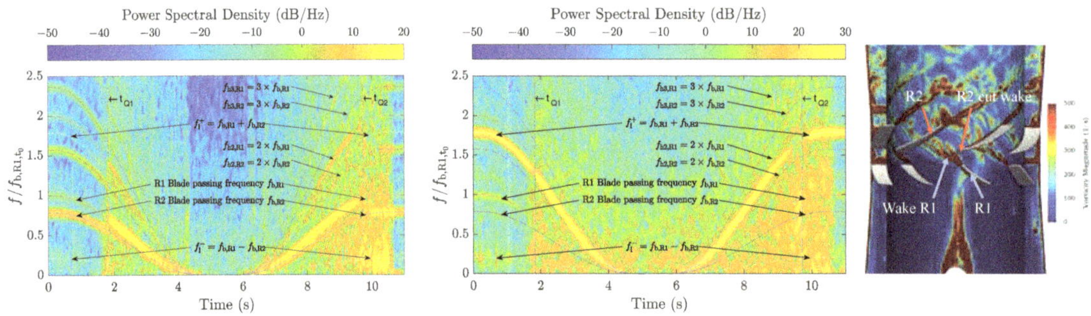

(**a**) STFT of pressure at probe P4. (**b**) STFT of force $F_{\mathrm{R1},x}$. (**c**) Vorticity.

Figure 13. Spectrogram of the STFT and vorticity magnitude at the design point, during the transient sequences in pump mode. The frequency is normalised with $f_{\mathrm{b,R1,t_0}}$, the blade passing frequency of Runner 1 at $\mathrm{t_0}$. Indices R1 and R2 denote Runner 1 and 2, respectively; b denotes the blade passing frequency; h2 is the second harmonic; h3 is the third harmonic; l is a linear combination. The vorticity is captured at the design point on a meridional plane and a cylinder (midpoint radius between hub and shroud) cutting the domain, close to the runners.

5. Conclusions

In this study, preliminary shutdown and startup sequences were numerically evaluated for a model counter-rotating pump-turbine in both pump and turbine modes. The rotational speed of each runner decreased from the design point to a stand-still, and increased back to the design point. As the rotational speed was close to zero, the flow field was highly unstable, and large flow structures were encountered. In turbine mode, the flow rate changed symmetrically with the rotational speed of the runners. In pump mode, reverse flow was captured as the machine lost the ability to build up sufficient head.

Hysteresis effects in the flow field were evident in the startup sequence in pump mode. This was because it took time for the flow to accelerate due to the changes in the rotational speed of the runners.

The rapid load variations that occur in low- or reversed flow conditions may lead to fatigue or damage the machine prematurely. The largest load variations occurred in pump mode, while they were less substantial in turbine mode. The shutdown and startup sequences need to be optimised in pump mode to minimise the loads on the runners. A valve should be added to reduce the flow rate to zero as the machine is stopped, and the operation of the valve should also be part of the optimisation procedure.

The main observed frequencies were related to the runners' blade passing frequencies. The strongest pulsations for a pressure probe located between the runners were in both pump and turbine mode, mostly correlated directly to the blade passing frequencies. In pump mode, it was found that the blade passing frequency of the downstream runner had a stronger power than the upstream runner. The positive linear combination of the blade passing frequencies showed the highest power for a transverse force component in pump mode. The harmonics of the main frequencies were also clearly visible through the transient sequences. A wide range of stochastic frequencies were excited as flow separation occurred by the blades of the runners, in both modes.

This study ultimately demonstrated that it is crucial to investigate transient sequences for the counter-rotating pump-turbine. One of the main reasons to use pumped hydro storage is to provide grid stability. To enable this in an effective manner, the knowledge of modelling and predicting the transient operations of the machine is of uttermost importance to ensure a high and stable production of renewable energy. The numerical framework presented in study demonstrated its functionality to predict and simulate transient shutdown and startup sequences for a novel counter-rotating pump-turbine. The suggested simulation strategy managed to capture some of the essential and expected flow phenomena, such as: vortex-shedding, flow separation, reversing the flow direction in pump mode, frequencies correlated with the runners, etc. It was hence concluded that the suggested numerical framework can be used to study the counter-rotating pump-turbine in time varying load conditions and transient sequences.

Author Contributions: Conceptualization, J.F.; methodology, J.F.; software, J.F., H.N. and S.S.; validation, J.F.; formal analysis, J.F. and S.S.; investigation, J.F.; resources, H.N.; data curation, n/a; writing—original draft preparation, J.F.; writing—review and editing, J.F., H.N. and S.S.; visualization, J.F.; supervision, H.N. and S.S.; project administration, J.F.; funding acquisition, H.N. All authors read and agreed to the published version of the manuscript. Please turn to the CRediT taxonomy (http://img.mdpi.org/data/contributor-role-instruction.pdf, accessed on 6 June 2021) for the term explanation.

Funding: This project received funding from the European Union's Horizon 2020 research and innovation programme under Grant Agreement No. 883553.

Institutional Review Board Statement: Not applicable.

Informed Consent Statement: Not applicable.

Data Availability Statement: Not applicable.

Acknowledgments: The computations were enabled by resources provided by the Swedish National Infrastructure for Computing (SNIC) at NSC and C3SE, partially funded by the Swedish Research Council through Grant Agreement No. 2018-05973. The blade geometries were designed and developed by Melvin Joseph and Mehrdad Zangeneh at Advanced Design Technology Ltd.

Conflicts of Interest: The authors declare no conflict of interest.

Abbreviations

The following abbreviations are used in this manuscript:

AMI	Arbitrary Mesh Interface
ALPHEUS	Augmenting Grid Stability Through Low Head Pumped Hydro Energy Utilization and Storage
CFD	Computational Fluid Dynamics
CRPT	Counter-Rotating Pump-Turbine
FFT	Fast Fourier Transform
LUST	Linear-Upwind Stabilised Transport
PHS	Pumped Hydro Storage
R1	Runner 1
R2	Runner 2
SAS	Scale-Adaptive Simulation
SST	Shear Stress Transport
STFT	Short-Time Fourier Transform

The following symbols are used in this manuscript:

ρ	Density	kg/m^3
η	Efficiency (hydraulic)	-
F	Force	N
f	Frequency	Hz
f_b	Frequency, blade passing	Hz
f_h2	Frequency, second harmonic ($2 \times f_\mathrm{b}$)	Hz
f_h3	Frequency, third harmonic ($3 \times f_\mathrm{b}$)	Hz
f_1^-	Frequency, linear combination (negative) of blade passing	Hz
f_1^+	Frequency, linear combination (positive) of blade passing	Hz
g	Gravity acceleration	m/s^2
H	Head	m
p	Pressure	Pa
Ω	Rotational speed	rad/s
n	Rotational speed	rpm
t	Time	s
t_end	Time, end of transient sequence	s
t_0	Time, start of transient sequence	s
t_Q	Time, zero flow rate	s
T	Torque	Nm
u	Velocity	m/s
Q	Volumetric flow	L/s (SI unit is m^3/s)

References

1. IEA. *Renewables 2020*; IEA: Paris, France, 2020. Available online: https://www.iea.org/reports/renewables-2020 (accessed on 6 June 2021).
2. IEA. *Will Pumped Storage Hydropower Expand More Quickly Than Stationary Battery Storage?* IEA: Paris, France, 2019. Available online: https://www.iea.org/articles/will-pumped-storage-hydropower-expand-more-quickly-than-stationary-battery-storage (accessed on 6 June 2021).
3. IHA. *2020 Hydropower Status*; IHA Central Office: London, UK, 2020.
4. Luo, X.; Wang, J.; Dooner, M.; Clarke, J. Overview of current development in electrical energy storage technologies and the application potential in power system operation. *Appl. Energy* **2015**, *137*. [CrossRef]
5. Rehman, S.; Al-Hadhrami, L.M.; Alam, M.M. Pumped hydro energy storage system: A technological review. *Renew. Sustain. Energy Rev.* **2015**, *44*. [CrossRef]
6. Deane, J.P.; Ó Gallachóir, B.P.; McKeogh, E.J. Techno-economic review of existing and new pumped hydro energy storage plant. *Renew. Sustain. Energy Rev.* **2010**, *14*, 1293–1302. [CrossRef]
7. ALPHEUS H2020. Available online: https://alpheus-h2020.eu/ (accessed on 26 April 2021).
8. Qudaih, M.; Engel, B.; Truijen, D.; Kooning, J.; Stockman, K.; Hoffstaedt, J.; Jarquin Laguna, A.; Ansorena Ruiz, R.; Goseberg, N.; Bricker, J.; et al. The Contribution of Low head Pumped Hydro Storage to a successful Energy Transition. In Proceedings of the Virtual 19th Wind Integration Workshop, Ljubljana, Slovenia, 11–13 November 2020.

9. Wintucky, W.T.; Stewart, W.L. Analysis of two-stage counterrotating turbine efficiencies in terms of work and speed requirements. In *Technical Report NACA-RM-E57L05*; NACA: Cleveland, OH, USA, 1958.
10. Carlton, J. *Marine Propellers and Propulsion*, 3rd ed.; Butterworth-Heinemann: Oxford, UK, 2012.
11. Furukawa, A.; Shigemitsu, T.; Watanabe, S. Performance test and flow measurement of contra-rotating axial flow pump. *J. Therm. Sci.* **2007**, *16*, 7–13. [CrossRef]
12. Kim, J.H.; Kasahara, R.; Kanemoto, T.; Miyaji, T.; Choi, Y.S.; Kim, J.H.; Lee, K.Y.; Galal, A.M. Multiobjective Optimization of a Counterrotating Type Pump-Turbine Unit Operated at Turbine Mode. *Adv. Mech. Eng.* **2014**, *6*. [CrossRef]
13. Momosaki, S.; Usami, S.; Watanabe, S.; Furukawa, A. Numerical simulation of internal flow in a contra-rotating axial flow pump. In *IOP Conference Series: Earth and Environmental Science*; IOP Publishing: Bristol, UK, 2010; Volume 12. [CrossRef]
14. Yang, Z.; Cheng, Y.; Xia, L.; Meng, W.; Liu, K.; Zhang, X. Evolutions of flow patterns and pressure fluctuations in a prototype pump-turbine during the runaway transient process after pump-trip. *Renew. Energy* **2020**, *152*. [CrossRef]
15. Xia, L.S.; Zhang, C.Z.; Li, H. Influences of runner blade shape on the transient behaviours of pump-turbines in load rejection. *J. Hydraul. Res.* **2020**. [CrossRef]
16. Fu, X.; Zuo, Z.; Chang, H.; Li, D.; Wang, H.; Wei, X. Mechanism of low frequency high amplitude pressure fluctuation in a pump-turbine during the load rejection process. *J. Hydraul. Res.* **2020**. [CrossRef]
17. Li, Z.; Bi, H.; Karney, B.; Wang, Z.; Yao, Z. Three-dimensional transient simulation of a prototype pump-turbine during normal turbine shutdown. *J. Hydraul. Res.* **2017**, *55*. [CrossRef]
18. Dixon, S.; Hall, C. *Fluid Mechanics and Thermodynamics of Turbomachinery*, 7th ed.; Butterworth-Heinemann: Oxford, UK, 2014. [CrossRef]
19. Melvin, J.; Fahlbeck, J.; Mehrdad, Z.; Nilsson, H. Optimized Initial Down-Scaled Model Design of SD-CRPT D2.1. Public V1.0, alpheus-h2020.eu, 2020. Available online: https://alpheus-h2020.eu/wp-content/uploads/2021/01/ALPHEUS_Deliverable_2.1_SD-CRPT_V1.0.pdf (accessed on 6 June 2021).
20. Fahlbeck, J.; Nilsson, H.; Salehi, S.; Mehrdad, Z.; Melvin, J. Numerical analysis of an initial design of a counter-rotating pump-turbine. In *IOP Conference Series: Earth and Environmental Science*; IOP Publishing: Bristol, UK, 2021.
21. Weller, H.G.; Tabor, G.; Jasak, H.; Fureby, C. A tensorial approach to computational continuum mechanics using object-oriented techniques. *Comput. Phys.* **1998**, *12*, 620. [CrossRef]
22. OpenCFD. *OpenFOAM—The Open Source CFD Toolbox—User's Guide, Version v1912*; OpenCFD Ltd.: Bracknell, UK, 2019.
23. Pope, S.B. *Turbulent Flows*; Cambridge University Press: Cambridge, UK, 2000. [CrossRef]
24. Versteeg, H.K.; Malalasekera, W. *An Introduction to Computational Fluid Dynamics: The Finite Volume Method*; Pearson Prentice Hall: Hoboken, NJ, USA, 2007.
25. Menter, F.R.; Egorov, Y. The Scale-Adaptive Simulation Method for Unsteady Turbulent Flow Predictions. Part 1: Theory and Model Description. *Flow Turbul. Combust.* **2010**, *85*, 113–138. [CrossRef]
26. Suh, J.W.; Yang, H.M.; Kim, J.H.; Joo, W.G.; Park, J.; Choi, Y.S. Unstable S-shaped characteristics of a pump-turbine unit in a lab-scale model. *Renew. Energy* **2021**. [CrossRef]
27. Farrell, P.; Maddison, J. Conservative interpolation between volume meshes by local Galerkin projection. *Comput. Methods Appl. Mech. Eng.* **2011**, *200*, 89–100. [CrossRef]
28. Aguerre, H.J.; Márquez Damián, S.; Gimenez, J.M.; Nigro, N.M. Conservative handling of arbitrary non-conformal interfaces using an efficient supermesh. *J. Comput. Phys.* **2017**, *335*, 21–49. [CrossRef]
29. Weller, H. Controlling the Computational Modes of the Arbitrarily Structured C Grid. *Mon. Weather. Rev.* **2012**, *140*, 3220–3234. [CrossRef]
30. OpenFOAM: User Guide v2006. Available online: https://www.openfoam.com/documentation/guides/latest/doc/ (accessed on 26 April 2021).
31. Uppström, L.; Fahlbeck, J.; Lillberg, E.; Nilsson, H. Simulation of a shut-down transient in the Francis-99 turbine model. In *IOP Conference Series: Earth and Environmental Science*; IOP Publishing: Bristol, UK, 2019; Volume 405. [CrossRef]
32. Fahlbeck, J. Implementation of an incompressible headLossPressure boundary condition. In *Proceedings of CFD with OpenSource Software*; Nilsson, H., Ed.; 2020. Available online: http://dx.doi.org/10.17196/OS_CFD#YEAR_2020 (accessed on 6 June 2021).
33. Savitzky, A.; Golay, M.J.E. Smoothing and Differentiation of Data by Simplified Least Squares Procedures. *Anal. Chem.* **1964**, *36*, 1627–1639. [CrossRef]
34. Kehtarnavaz, N. *Digital Signal Processing System Design*; Academic Press: Burlington, MA, USA, 2008. [CrossRef]
35. Lengani, D.; Santner, C.; Spataro, R.; Göttlich, E. Analysis tools for the unsteady interactions in a counter-rotating two-spool turbine rig. *Exp. Therm. Fluid Sci.* **2012**, *42*, 248–257. [CrossRef]

energies

Case Report

Community-Based Business on Small Hydropower (SHP) in Rural Japan: A Case Study on a Community Owned SHP Model of Ohito Agricultural Cooperative

Zafar Alam [1], Yoshinobu Watanabe [1], Shazia Hanif [2], Tatsuro Sato [3] and Tokihiko Fujimoto [4,*]

1 Nakayama Iron Works Co. Ltd., Saga 843-0001, Japan; zafar_a@nakayamairon.co.jp (Z.A.); watanabe@nakayamairon.co.jp (Y.W.)
2 Department of Agricultural Engineering, Muhammad Nawaz Sharif University of Agriculture, Multan 66000, Pakistan; shazia.hanif@mnsuam.edu.pk
3 Faculty of Architecture and Civil Engineering, Kyushu Sangyo University, Fukuoka 813-8503, Japan; sato@ip.kyusan-u.ac.jp
4 School of Political Science and Economics, Meiji University, Tokyo 101-8301, Japan
* Correspondence: fujimoto@meiji.ac.jp

Abstract: Energy is the prerequisite for social and economic development of a community and country. In Japan, national government is promoting small hydropower (SHP) through a renewable energy policy by providing a high FIT price of 34 yen (≒0.32 US$/kWh) on energy generated from an SHP of less than 200 kW. Until now, the energy generation was controlled by national government agencies, but now independent power generation businesses are growing at the local community level in rural Japan. For the future growth of SHP, it is necessary to make electricity generation at the local community level. Therefore, these local communities will install and manage their renewable electricity by themselves. It will help to make the community self-sustainable and independent from the national government, and at the same time, it will also lead them to achieve the Sustainable Developments Goals (SDGs) target from community-based action. This paper aimed to discuss an SHP development business model in which local community will become the business owner of the SHP. It means "of the community, by the community and for the community". The community identifies their renewable energy potential and needs, they borrow money from the financial organization or banks, install the power plant and do necessary maintenance and management by themselves. The revenue earned by selling electricity is used to repay the loan, and the rest is used for community development directly (such as local roads construction, agriculture land improvements, community hall maintenance, waterways maintenance, welfare, etc.). This paper also discussed a community-based 50 kW SHP installed in Miyazaki prefecture of Japan as a case study. This SHP is one of the best examples of a community ownership model (Community-based business model). A detailed explanation from planning to investment has been discussed. The local community is getting approximately 112,000 USD per year by selling the electricity, and 162-ton CO_2 is estimated to decrease yearly, which will support the achievement of SDGs. Finally, installing this kind of SHP in remote areas will provide managerial skills to the local community directly, plant operation knowledge, and education to local students. Local communities learn the problem-solving skills, which lead them to solve the local problem on a community level by themselves.

Keywords: community development; community ownership; small hydropower; SHP; renewable energy; crowdfunding; FIT; community-based business; agricultural cooperative

Citation: Alam, Z.; Watanabe, Y.; Hanif, S.; Sato, T.; Fujimoto, T. Community-Based Business on Small Hydropower (SHP) in Rural Japan: A Case Study on a Community Owned SHP Model of Ohito Agricultural Cooperative. *Energies* **2021**, *14*, 3349. https://doi.org/10.3390/en14113349

Academic Editors: John M. Cimbala and Bryan J. Lewis

Received: 14 May 2021
Accepted: 4 June 2021
Published: 7 June 2021

1. Introduction

In the present scenario, increasing world energy demand and climate change imperatives drive the dependency of energy sources towards available renewable energy resources. Climate change is one of the major factors that has a strong impetus on the way of thinking about world energy dependency. It urges the world to achieve alternative renewable resources other than its reliance on fossil fuels. In this regard, it is not only requisite to use all the possible renewable resources, but also to shift the established fossil fuel-based energy system toward renewable energy. Therefore, it is necessary to exploit domestic energy sources, especially renewable energy sources, as their potential is high.

Moreover, for UN- Sustainable Development Goals (SDGs), it is required for a country to devise policies and form strategies to increase the share of renewable energy in the total energy mix, and ensure the provision of electricity to the rural and remote areas. In rural areas, there are a variety of renewable energy options available, e.g., hydropower, solar, wind and biomass. Many mathematical models have been used to make decisions among all available renewable energy resources. Geological position and terrain are characterized as strong impact for the assessment of most suitable renewable energy resources, e.g., in Brazil, wind energy is considered the most favorable renewable energy source [1]. In addition, environmental and human health impacts are also very important consideration for the selection [2].

Being a target part of SDGs, sustainable supply chain management is prerequisite for the modern world in order to conserve the available resources, reduce the waste as well as generate revenue. It not only helps the natural environment, but it can also give enterprises financial benefits [3]. In addition, their efforts can be more effective if integrated approach and regional networking are involved [4]. Many researches highlight the constraints and barriers related to the implementation of SDGs in supply chains. For sustainable development, organizational management is responsible for dealing with social, economic as well as ecological performance at the same time. However, sometimes implementation of the goals is associated with enormous problems, which deject the continuity of work and ultimately discourage business managers [5]. The purpose of renewable energy is not only to raise the rural economy, but also to save the ecological environment [6]. The main objective of sustainable development is to provide better life to human beings by wise utilization of the natural resources and considering its limitations on the one Earth [7].

Small hydropower (SHP) is considered a reliable and sustainable source for making electricity. As a potential future source of energy, it has become an increasingly attractive choice, which plays an important role for its adoption and development all over the world. It is considered environmentally friendly, as its construction has small effects on water bodies. SHP gain more preference because of its available potential, small scale construction requires less investment, speedy construction, low cost generate quick revenue as well as environmentally friendly clean energy [8,9]. The other benefits are to reduce the safety risks associated with small dams, and lesser population displacement or land usage issues [10]. Small hydropower does not require a large storage structure like dam construction as compared to large scale hydropower, thus reducing the technology's environmental impacts [11]. For SHP, potential power available, head, and flow considerations are very important constraints regarding the selection of site-specific parameters [12]. Sites with a reliable water supply year-round and a large vertical drop in a short distance are considered as the best locations for the installation of SHP.

In Japan, hydroelectricity is considered the second most important renewable energy source after solar energy. There are several reports regarding the economic benefits of implementing SHP, considering ecological sustainability and environmental ecosystem, and their impacts [13–16]. Overall, Japan has an abundance of small streams particularly in mountainous regions and in the past many small channels have been created for irrigation purposes. Some of these irrigation channels continue to be used today. Others, though in good condition, remain unused. Increasingly, these channels are being utilized to produce SHP electricity to promote local development. This is particularly important for rural Japan, which has suffered from rapid depopulation due to a combined effect of aging population and urban migration [17].

Small hydropower has regained attention in Japan since the 2000s, when climate change and climate crisis were highlighted. In 2003, the "Act on Measures Concerning the

Use of New Energy by Electric Power Companies" (RPS Act) targeted small hydropower of 1000 kW or less, so that about 10 new construction works will be carried out annually. The RPS Law was taken over by the "Act on Special Measures Concerning the Procurement of Renewable Energy Electricity by Electricity Companies" (commonly known as the FIT Law) in 2012, and under the feed-in tariff (FIT); in recent years, the small hydropower market has been gradually revitalizing. The Japanese government has started to support the SHP project in order to support a renewable policy in Japan as well as to prevent migration and to protect the old rich culture of the respective communities [18].

To promote the SHP, the Japanese government is giving a high FIT of 34 yen per kWh (≒0.32 US$/kWh) under the capacity of 200 kW, which enables a faster payback [18,19]. In spite of FITs and other benefits, planning a new SHP plant, searching for the initial investment, consensus building among the community people and various stakeholders, convincing local people is not an easy task and so, and despite having hydro potential in a number of communities, it has not been harnessed yet. This paper will discuss an ongoing business model idea to develop community-based SHP in rural areas, which is based on:

- Of the community;
- By the community;
- For the community.

In Japan, the concept of community development by SHP installation is not new. A number of SHP were installed in 1940–1970. In Japanese remote areas, many SHP run by local agricultural cooperatives (hereinafter referred to as agricultural cooperatives) exist in the Chugoku region in the west part of the main island in Japan. By the end of March 1955, about 90 SHP operated by agricultural cooperatives and others were built in the Chugoku region, while 181 were constructed nationwide. To this day, some facilities were abolished or suspended due to disaster, aging, poor management, submergence and other reasons. By 1980, 74 remained, and currently 54 are still generating. All 54 facilities currently in operation are less than 1000 kW, with a maximum of 660 kW, a minimum of 24 kW, and an average of 189.1 kW. The total installed capacity of the 54 facilities is 10,209 kW [20,21].

The purpose of this study is to provide a SHP development business model in which local communities come forward for organizing and development of SHP and become the business owner. This paper consists of three sections: the very first section explains a community owned SHP development model. It gives an idea of community ownership to develop an SHP. This section also discusses the project financing, multi-stoke holder, crowd funding, etc., with regard to SHP development. It provides an example of community development that will work collaboratively for common goals. It is expected that community level SHP development will lead to many more SHP. The second section of this paper is analyzing a case study of a community owned SHP installed in rural Japan. This section explains techno-economic details of installed SHP and its social and economic impact on community and environmental impact on climate change. The third section discusses its effects, followed by the conclusion and limitations.

2. Community Owned SHP Development Model

This section explains community development by installing SHP, as a community-based business by local people.

In this model, the local residents will become business owners, it creates jobs locally, and it motivates young people to work in their hometown. Additionally, it will attract the migrated peoples from urban areas to do business in their hometown again (called U turn). The installation of this kind of project intends to solve local problems by utilization of local resources. It aims to build a better relationship between urban consumers and energy production areas [22].

Step1—The identification of SHP potential, feasibility study of the potential sites must be done by the local government or prefectural government with help from a consultant, developer or NGOs of SHP. There should be consensus building between the local community, local and prefectural government and the developer, consultant or NGOs. The consultant should explain the benefits of installing this project to the local community by organizing seminars and consensus building meetings. In most of the cases, the local community people wish to install SHP projects, but they have some concerns about project execution, management and investment. Community people need moral support and training for fundraising, management and operation.

Step2—It is about finding low interest rate investors. In Japan, for an individual person, getting a loan from the bank or any finance is difficult. Additionally, the interest rate is quite high. However, if a community corporation needs a loan for communal use, they can get it easily. Moreover, if the corporation or company comprises farmers, then the interest rate is very low due to its social impact. To avail this opportunity, a local cooperatives and community-based business company are made, and the local people become the owner and members of this community-based business association. This new cooperative association gets the loan from the bank or financial institution and becomes the SHP project owner. In the future, this association will pay the loan, do the necessary maintenance, and utilize the revenue for the community development directly. All members of this association will be the owner of the SHP project at the community.

Step3—Community-based business model (community ownership, multistate holders, networking and various responsibilities) represents the network in which cooperative associations collaborate with the stakeholders, construction companies, consultants, mechanical and electrical makers of SHP. Cooperatives comprise individuals with mutual resources to provide input, output, and capital resources to members at a low cost.

Figure 1 represents the business flow for community SHP development projects. This networking strives to provide information and support for sustainable agriculture and community development. It develops mechanisms to monitor, promote public–private partnerships (PPP), improve smallholder agriculture, and reduce agricultural losses. Funding comes from voluntary contribution, development banks, and other local resources.

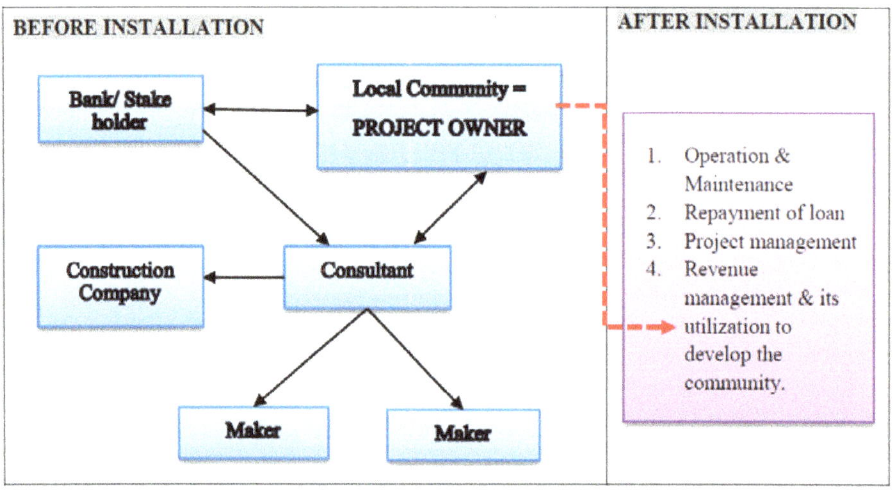

Figure 1. Community-based business model of SHP; stakeholder assigned before and after the project installation (Source: Author).

Figure 2 represents the responsibilities of each stakeholder toward development and execution of the project. It shows the strong initiative and struggle done by the local community for self-sufficiency in energy production. It also encourages the community to manage its available renewable energy resources along with local commons [23].

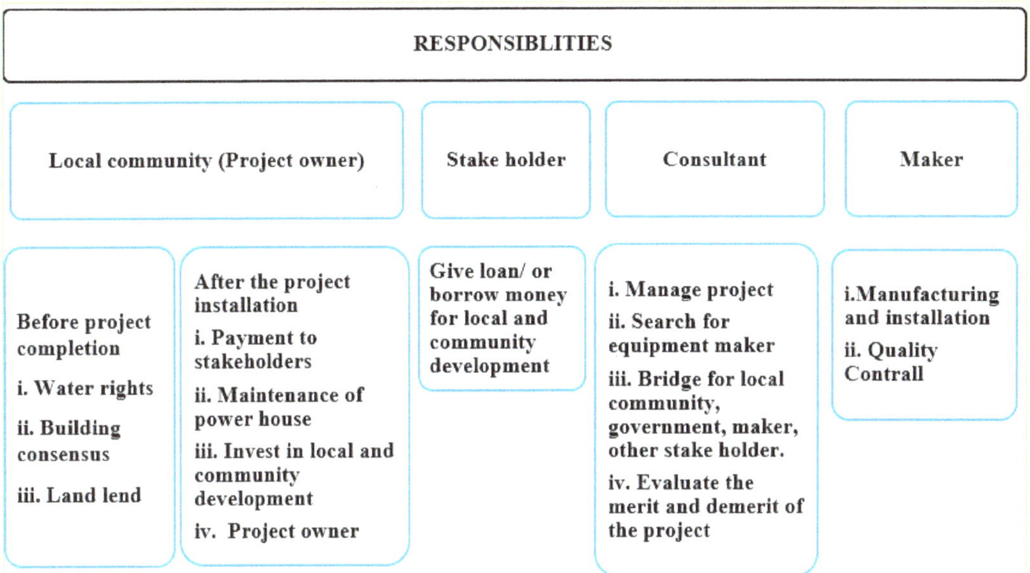

Figure 2. Responsibilities of each stakeholder of the project (Source: Author).

In this kind of SHP development project initiated by the community, it is necessary to utilize the local resources as much as possible. Generally, local civil companies do the civil engineering parts, locally available construction material would be used to make the powerhouse and local labor. This helps to boost the local economy and job opportunities to local people.

3. A Case Study of 50 kW Ohito SHP Installed in Miyazki Prefecture, Japan

This section discusses with the case study of the development of 50 kW Ohito small hydropower plant. This SHP has been installed in Ohito agricultural cooperatives owned by residents. Figure 3 shows the location map of Ohito SHP, which comes under Hinokage town in Miyazaki Prefecture. As per 1 October 2019 data, the estimated population of Ohito community is 3656 people (men and women). The total field area is 277.67 km^2. It is located next to the historically significant town of Takachiho. Hinokage town is known primarily for its three large modern bridges, traditional Kagura dances, and the work of Kazuo Hiroshima, a traditional weaver of bamboo baskets.

Like many districts in rural Japan, the area has been struggling with an aging population and an exodus of young people to Tokyo and other major cities in Japan. The Hinokage town's population peaked in the 1950s at 16,199 people, but has steadily declined since the closure of the Mitate mine in March 1970. Between 2000 and 2005, four area elementary and middle schools were closed, and more closures are planned very soon.

Figure 3. Location map (Source: Google map).

3.1. *Ohito SHP Development Model (A Community Ownership Concept)*

The Ohito SHP project has been started with the participation and collaboration of the local people inside the community. Firstly, the Ohito community identified the problems through which they were suffering (for example, depopulation, unavailability of local jobs, inaccessibility of electricity and other facilities and detachment from the big city). They wanted to utilize the extra water of an existing irrigation in the community, which is used for irrigation of nearby crops. The main crops are paddy, which uses a high amount of water for 4 months in a year, but for the remaining 8 months, the water was wasted at the time.

To build SHP, all the community residents jointly made a community-based business company named "Ohito Agricultural Cooperatives". There are 53 households in the Ohito community, each household becoming a member of this newly established association.

Figure 4 represents the formation structure and contribution of Ohito Agricultural Cooperatives for installing Ohito SHP. All households of the Ohito community contributed a capital amount of 4000 JPN yen (≒38 USD) to establish Ohito Agricultural Cooperatives. It tries to take care of all of the community problems of mutual concerns regarding water rights and growing of a new variety of crops. It makes the village not only self-sufficient in electricity demand, but also provides handsome annual earnings. This co-operative takes all the responsibilities before and after the execution of the SHP project; maintenance of the water way and powerhouse, loan instalment payment and water use cost. The major responsibility of the organization is community development, management of financial resources, capacity building and management of the facility.

Ohito Agricultural Cooperatives uses the expertise of staff and other professionals to execute the project and data that aid development. This mobilizes and manages the funds, loans and subsidies provided by local residents, Miyazaki prefecture and banks to make sure the projects achieve their goals, respectively. It is also included in the responsibilities of the co-operatives to protect rural livelihoods, advance farming and uplift the living standards of locals.

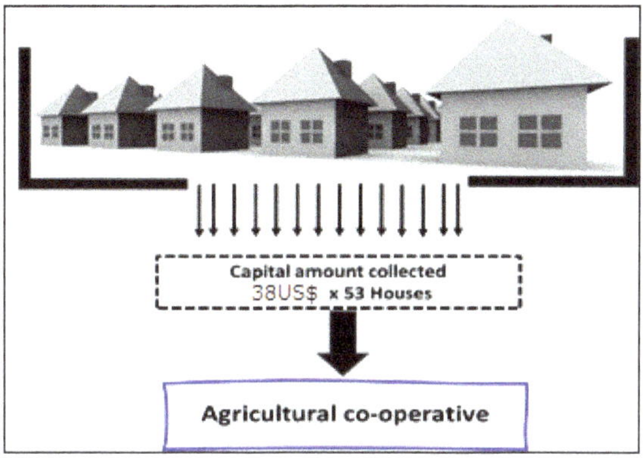

Figure 4. Formation of Ohito Agricultural Cooperatives (Source: Author).

3.2. Technical Specification of Ohito SHP

Total installed capacity of Ohito SHP is 50 kW and an annual power generation of approximately 320 MWh/year was estimated. This SHP plant contains an irrigation channel/intake/forebay tank. Figure 5 shows the powerhouse of Ohito SHP. The powerhouse is made up of locally available stone to save money and involve the local resources. The generated electricity is sold to the national grid and it is connected to 200 V poles available near the powerhouse. The total length of this channel is 10 km. At the middle of this channel (approx. 5 km), a small desilting tank with a tyrolean type intake weir has been made as shown in Figure 6. The water was first stored in a forebay tank. Since the priority is given to irrigation. The rest water is utilized to produce electricity, and so the amount of discharge changes throughout the season according to water requirements for the crops.

For this site, a cross flow turbine (made in Indonesia by Asosasi Hidro Bandung) has been selected with electrical equipment and a control panel (made in Japan). A high efficiency IPM generator with AC-DC converter (GD) and further on DC-AC converter (GC) have been installed as shown in Figure 7. It is a unique system that increases the system efficiency even on low output and low speed of the generator. An IPM generator is a high efficiency (~95%) varying speed generator, which is approx. 80% efficient at 25% of its rated speed. The combination of GD and GC helps to stabilize the output voltage and frequency according to the grid requirement.

Figure 5. Power house (Source: Author).

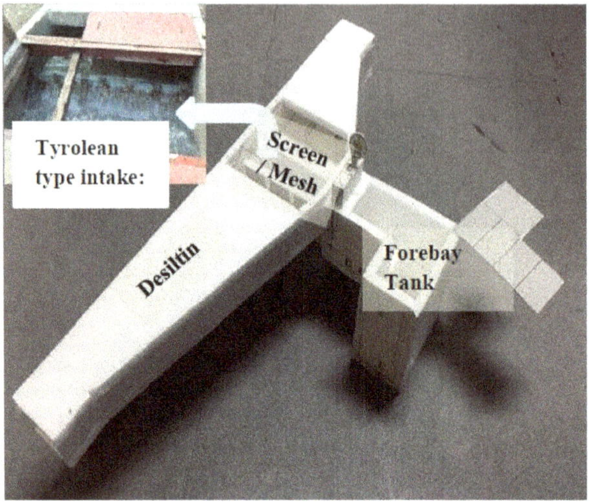

Figure 6. Tyrolean intake with forebay tank (Source: Author).

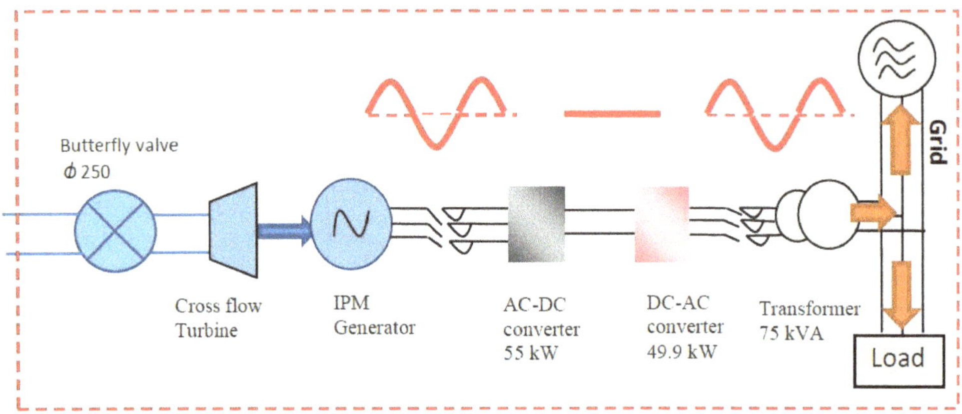

Figure 7. System flow of Ohito SHP (Source: Author).

Table 1 presents the site details and equipment specification installed at 50 kW Ohito SHP. The low cost overseas turbine makes the system cheaper, and a high efficiency electrical system, especially on low load condition, is an attractive concept for other similar community development projects. In a case of SHP construction in Japan, one of the major concerns is a garbage removal process. Due to low capacity, it is not economically beneficial to use an automatic garbage removal system by Tyrolean intake. Another obstacle arises due to powerhouse location (generally in mountainous regions away from the downtown). Someone must go and clean the garbage daily, it costs time, money and effort for garbage removal activities. Keeping that in mind, a Tyrolean type automatic garbage cleaning intake has been constructed.

Table 1. Technical details of the 50 kW Ohito SHP (Source: Author).

Site Condition and Parameters	
Effective head	85 m
Design discharge	0.12 m^3/sec
Turbine	Cross-flow T15 (D300, B100) from Asosiasi Hidro Bandung (Indonesia)
Efficiency of turbine	76%
Type of generator	IPM generator from Yasukawa Electronics (Fukuoka, Japan)
Generator specification	1150 rpm, 55 kW Generator + GD + DC efficiency = ~94/95/95%
Consultant	River Village Co. Ltd. (Fukuoka, Japan)
Mechanical & electrical design concept	Nakayama Iron Works Co. Ltd. (Saga, Japan)
Civil Engineering	Local constructor (Ohito, Miyazaki, Japan)

3.3. Project Financing & Economic Parameters

The Ohito Agricultural Cooperatives borrowed the loan from financing cooperation and banks. At the same time, they got a subsidy from the Miyazaki prefectural government. Figure 8 represents the construction cost and finance arrangement for the construction. The total cost of construction was USD 896,000. Out of which, USD 94,340 is subsidized by the government of Miyazaki prefecture. The others are arranged by the loan from the local bank of Miyazaki and Japanese Financing Corporation at the rate of interest 2.6% and 0.9%, respectively, within the repay period of 20 year, which is guaranteed by FIT.

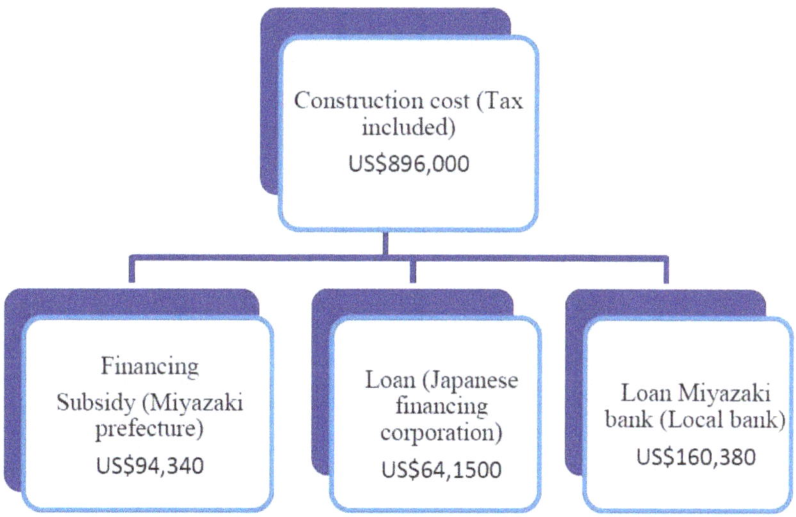

Figure 8. Business model and finance distribution for the construction of Ohito SHP (Source: Author).

Figure 9 represents a cumulative cash flow estimation of Ohito SHP. The running cost was calculated as approximately USD 8500 per year. The results of economic calculation are as follows: Project IRR (4.3%), B/C ratio (1.36) and project payback period (13 years). In Figure 8, the *x*-axis shows the period. Since the FIT provided by the Japanese government ended for 20 years. The cost analysis was done for a 20 years period. The *Y*-axis gives the cumulative cost per year, which includes total earning−maintenance cost−loan payment per year. The breakeven will be achieved after 13 years by the calculation.

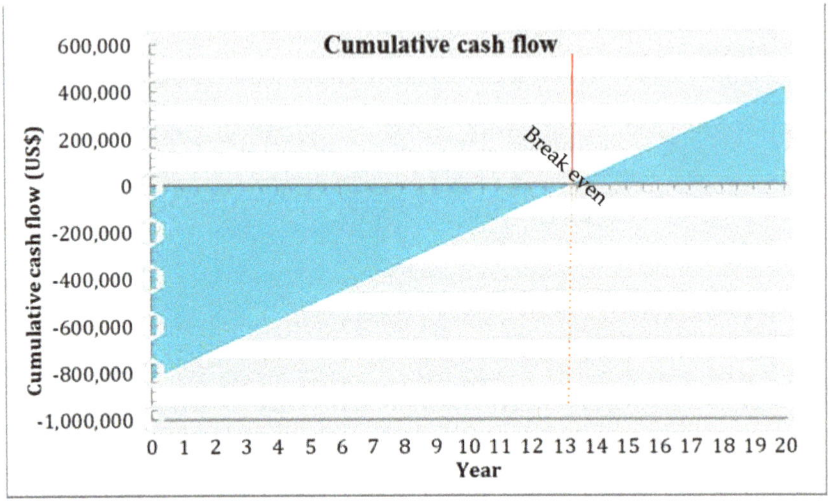

Figure 9. Cumulative cash flow of Ohito SHP (Source: Author).

4. Discussion

Installing SHP has several recognizable benefits, including in achieving SDGs from community-based action, and helping in the development of local community people or rural areas, which have limited jobs and facilities.

More precisely, it supports the growth of rural communities and local developers. In the case of Ohito SHP, a local civil engineering company performed all the civil works, local traditional stones and wood were used to build the powerhouse. Furthermore, community people get the benefit by selling electricity to the Kyushu Electric Power Co., Inc., Fukuoka city, Japan (one of national electric power companies), and local companies get work orders, which helps them. As discussed above, the Japanese government gives high FIT by a renewable energy policy, and the FIT provided on SHP of less than 200 kW is 34 yen (US$ 0.32 per kWh) for 20 years, which means an SHP developer gets high earnings by selling electricity generated from SHP.

From the economic point of view, Ohito SHP generates more than 350 MWh per year. It continuously runs at its full capacity of 50 kW, except for a few months during rice crop harvesting (4 months in a year), during that, capacity reduces to 20 kW. The plant started in January 2018 and within the last approximately 3 years, 1100 MWh energy has been generated as per the data seen on 3 March 2021. The total revenue earned until the mentioned date was = 1100 × 1000 × 0.32 (FIT price) = 352,000 US$. On average, 350 MWh energy has been generated in the last 3 years. If the plant runs similarly as now, it is possible that the developer will achieve the breakeven point sooner than estimated.

The energy output and power generation for the last month of 10 April 2021 to 10 May 2021 is shown in Figures 10 and 11. It shows the variation of power output and power generation with the time. It shows that power output and energy generation was constant until 2 May 2021, and then afterwards, it reduced to 1/3 of designed output. Generator power and power going to the grid with respect to time and water pressure have been shown in Figures 9 and 10. It has been seen that until 2 May 2021, the power output was 50 kW approximately, but after this date, the output decreased from 50 kW to around 17 kW. It is because the water is being used for irrigation purposes, which has priority over power generation. This shows a new community-based business of the Ohito community integrated in agriculture and hydroelectric generation at the local level.

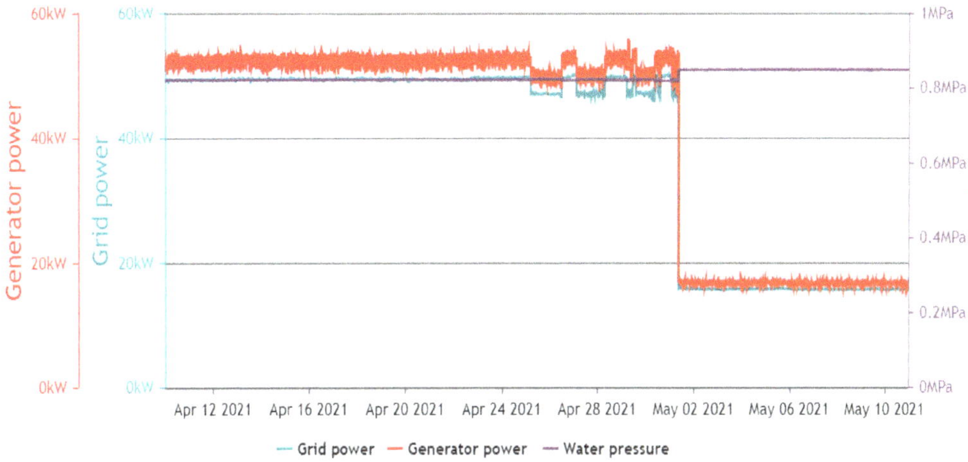

Figure 10. Power generation and supply to grid (Source: Author).

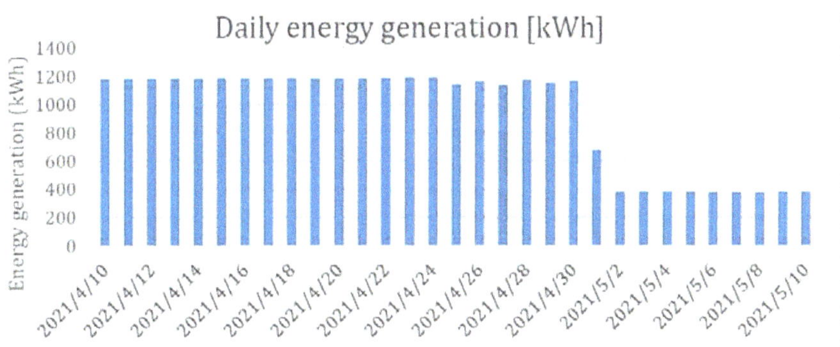

Figure 11. Energy generation per day in kWh. Approximately 1 MWh energy was being generated until 2 May 2021, which has decreased to 400 kWh after starting agricultural activities. The water is used for rice crop irrigation (Source: Author).

From the environmental point of view, the SHP development supports the government's efforts to promote clean energy sources and to carry out its target of SDGs, as an SHP uses water to produce electricity without consuming it. The Ohito SHP uses the existing wastewater in the irrigation channel to produce useful clean renewable energy. In terms of specific SDG, Ohito SHP supports targets 7, 8, 13. With respect to CO_2 reduction, since, per kWh of electricity generated from SHP reduces 0.463 kg CO_2, the 350 MWh of electricity generated annually by the Ohito SHP will lead to reduce approximately 162 tons of CO_2 per year.

In terms of social contribution, a community-based SHP leads the local growth by providing work to local companies, utilizing local resources and providing a means of engagement to the local young people and a learning opportunity to the students in the community. Like other communities in rural Japan, the Ohito community is also facing ageing, lack of job opportunities, depopulation, etc. Installing this kind of SHP will provide local empowerment and revival of local business. The sole idea to install the Ohito SHP was as follows:

- Improving social welfare, especially improving the living support system.
- Maintenance of roads, irrigations, waterways, inside the community.

- Maintenance of public halls.
- Utilization of degraded farmland aiming for bamboo grove maintenance and agricultural industry.

However, unfortunately, there was no such work done until the writing of this paper. All the money (~35,200 USD) is still being kept by the local community. No such investment has been made. Therefore, the direct social impact of installing this plant has not started yet, but there were some indirect social impacts in terms of idea sharing and attracting other communities to install similar SHP projects in respective communities. In addition, it is a source of education for the community people and university students, as they learn not from theory, but by looking at the actual running plant. There is one more hidden important impact. The Ohito community was getting implicit know-hows and community experiences "of the local community, by the community, for the community", by actually operating SHP.

5. Conclusions

This paper discussed an SHP development business model in which the local community will become the business owner. It is focused "of the community, by the community and for the community" by discussed a case study of community-based 50 kW SHP installed in rural Japan. This SHP is one of the best cases of a community ownership SHP model in Japan. The local community is getting approximately 112,000 USD per year by selling the electricity and 162 tons CO_2 are estimated to decrease yearly, which will support the achievement of SDGs. Installing this kind of SHP will provide managerial skills to the community, plant operation knowledge, and education to local people. Local communities learn the problem-solving skills and implicit know-how, which lead them to solve the local problem on a community level. Finally, it is concluded that SHP can play a vital role in community development. SHP makes a network of stakeholders, it integrates the local community to the urban stakeholders, and it creates jobs for the local community, which help to engage the local youth to be at home.

In recent times, a localized, independent generation of energy has started and it is growing. An SHP achieves break-even within 8–12 years depending on the investment, but the Japanese government gives FIT for 20 years. This money helps local people to do their community development. The community no longer depends on government subsidies for their development from now on. They use this SHP money for their agriculture land improvement, renovation of local community halls, roads, etc. It gives a means of independence to the local community.

To install more and more SHP, there is a need to make an attractive community-based business model, which includes both rural people and urban stakeholders. This paper has suggested a community owned SHP model, which can be used to install more and more SHP for the local community. It gives ownership to the local community, as the local people identify their problems, install the SHP and do necessary maintenance and management. The income generated by selling electricity is received by the newly established corporation/company composed of all the community members. This association pays the loan to finance companies or banks and utilizes the rest of this money for the development locally. This new community-based action not only works on SHP development, but also the planning to use other natural resources available in the community. It will promote the community independence and self-sustainable to make sustainability.

With a number of benefits, there are some limitations that need to be fixed in the future.

First, the Japanese government is paying a high FIT on SHP, but it still needs to revise. The equipment's and labor costs are high, and so the rate of return becomes slow. If 100% money is borrowed, the project payback period increases to 10–12 years, which should be below 10 years.

Second, local communities are not always aware and do not have much skill and confidence to manage and operate their natural resources. Therefore, they do not decide to invest the revenue earned by selling electricity. In the future, it is necessary to make a plan for the proper utilization of sales revenue. It is necessary to introduce a new player who will take care of this investment to develop the local culture, agriculture land, and community needs. There is a possibility to include university professors or NGOs or the local city office as a coordinator and interpreter. They will take care of revenue earned to use properly for the development of community road, culture, agriculture needs, etc. The community people will decide about the investment area and they will inform them.

Author Contributions: Conceptualization, Z.A. and T.F.; methodology, Z.A.; software, S.H.; validation, T.F., S.H. and Y.W.; formal analysis, S.H.; investigation, Z.A.; resources, T.F.; data curation, Z.A., T.F. and Y.W.; writing—original draft preparation, Z.A. and S.H.; writing—review and editing, T.F. and S.H.; visualization, Y.W.; supervision, T.F. and T.S.; project administration, T.S. and Y.W.; funding acquisition, T.F. All authors have read and agreed to the published version of the manuscript.

Funding: JSPS KAKEN, 18K14538, Tokihiko Fujimoto.

Institutional Review Board Statement: Not applicable.

Informed Consent Statement: Not applicable.

Data Availability Statement: Not applicable.

Conflicts of Interest: The authors declare no conflict of interest.

References

1. Da Silva, C.G. Renewable energies: Choosing the best options. *Energy* **2010**, *35*, 3179–3193. [CrossRef]
2. Neri, E.; Cespi, D.; Setti, L.; Gombi, E.; Bernardi, E.; Vassura, I.; Passarini, F. Biomass Residues to Renewable Energy: A Life Cycle Perspective Applied at a Local Scale. *Energies* **2016**, *9*, 922. [CrossRef]
3. Hondo, H. Life cycle GHG emission analysis of power generation systems: Japanese case. *Energy* **2005**, *30*, 2042–2056. [CrossRef]
4. Zimon, D.; Tyan, J.; Sroufe, R. Drivers of sustainable supply chain management: Practices to alignment with un sustainable development goals. *Int. J. Qual. Res.* **2020**, *14*, 219–236. [CrossRef]
5. Koirala, B.P.; Vaghela, D.; Mitavachan, H.; Kulenthran, R. Opportunities and Challenges of Community Energy Systems: Analysis of Community Micro-hydro Systems in South and South-East Asia (SSEA). In Proceedings of the MES-BREG: Innovating Energy Access for Remote Areas: Discovering Untapped Resources At: University of California, Berkeley, Berkeley, CA, USA, 10–12 April 2014. [CrossRef]
6. Kong, Y.; Wang, J.; Kong, Z.; Song, F.; Liu, Z.; Wei, C. Small hydropower in China: The survey and sustainable future. *Renew. Sustain. Energy Rev.* **2015**, *48*, 425–433. [CrossRef]
7. Fonseca, L.M.; Domingues, J.P.; Dima, A.M. Mapping the Sustainable Development Goals Relationships. *Sustainability* **2020**, *12*, 3359. [CrossRef]
8. Cheng, H. Utilization of small hydropower resources with promotion of new countryside construction. *China Water Resour.* **2006**, *14*, 17–18.
9. Paish, O. Small hydro power: Technology and current status. *Renew. Sustain. Energy Rev.* **2002**, *6*, 537–556. [CrossRef]
10. Egréa, D.; Milewskib, J.C. The diversity of hydropower projects. *Energy Policy* **2002**, *30*, 1225–1230. [CrossRef]
11. Hansen, M.; Simmons, R.T.; Yonk, R.M. The Regulatory Noose: Logan City's Adventures in Micro-Hydropower. *Energies* **2016**, *9*, 482. [CrossRef]
12. Kosnik, L. The potential for small scale hydropower development in the US. *Energy Policy* **2010**, *38*, 5512–5519. [CrossRef]
13. Gorla, L.; Perona, P. On quantifying ecologically sustainable flow releases in a diverted river reach. *J. Hydrol.* **2013**, *489*, 98–107. [CrossRef]
14. Dee, N.; Baker, J.; Drobny, N.; Duke, K.; Whitman, I.; Fahringer, D. An environmental evaluation system for water resource planning. *Water Resour. Res.* **1973**, *9*, 523–535. [CrossRef]
15. Botelho, A.; Ferreira, P.; Lima, F.; Pinto, L.M.C.; Sousa, S. Assessment of the environmental impacts associated with hydropower. *Renew. Sustain. Energy Rev.* **2017**, *70*, 896–904. [CrossRef]
16. Schramm, M.P.; Bevelhimer, M.S.; DeRolph, C.R. A synthesis of environmental and recreational mitigation requirements at hydropower projects in the United States. *Environ. Sci. Policy* **2016**, *61*, 87–96. [CrossRef]
17. Watanabe, Y.; Alam, Z.R.R.; Okajima, K.; Yamamoto, S.; Faisal, R.; Kumar, A.; Fujimoto, T. *World Small Hydropower Development Report 2019 case studies*; United Nations Industrial Development Organization (UNIDO) and International Center on Small Hydro Power (ICSHP): Viena, Austria; Hangzhou, China, 2019; pp. 37–40.
18. Japan's Ministry of Economy, Trade, and Industry. *FY2019 Annual Report on Energy (Energy White Paper 2020). Outline. October 2020 and Fifth Strategic Energy Plan*; Japan's Ministry of Economy, Trade, and Industry: Tokyo, Japan, 2020.
19. Japan's Ministry of Economy, Trade, and Industry. *FY2011 Annual Report on Energy (Energy White Paper 2012) Outline. October 2011*; Japan's Ministry of Economy, Trade, and Industry: Tokyo, Japan, 2012.

20. Fujimoto, T.; Minata, K.; Shimatani, Y. Chugoku Chihou no Syousuiryoku Energy riyou ni miru Shizen Energy ni motozuku Chiiki zukuri no Shisou (In Japanese). *Shimaneken Chuusankan Chiiki Kenkyu Center Kenkyu Houkoku* **2012**, *8*, 31–38.
21. Fujimoto, T. *World Small Hydropower Development Report 2016*; United Nations Industrial Development Organization (UNIDO) and International Center on Small Hydro Power (ICSHP): Viena, Austria; Hangzhou, China, 2016; pp. 415–419.
22. Fujimoto, T.; Kagohashi, K. Community-ledMicro-Hydropower Development Guided by Landcare Approach: A Case Study on a Networking activities of local residents and farmers in Gokase Town, Japan. *Energies* **2019**, *12*, 1033. [CrossRef]
23. Data Max Netib-News. Article on Future of Local Communities Connected by Small Hydropower, Published on 05 March 2018. 2018. Available online: https://www.data-max.co.jp/article/2985?rct=machi7 (accessed on 25 April 2021).

MDPI
St. Alban-Anlage 66
4052 Basel
Switzerland
Tel. +41 61 683 77 34
Fax +41 61 302 89 18
www.mdpi.com

Energies Editorial Office
E-mail: energies@mdpi.com
www.mdpi.com/journal/energies